EVOLUTIONARY BIOLOGY OF THE PRIMATES: An Introductory Reader

Daryl G. Frazetti

University of Massachusetts Boston - May 2002 (2004)

"There is something fascinating about science. One gets such wholesale returns of conjecture out of such a trifling investment of fact".

- Mark Twain, Life on the Mississippi, 1874

Description:

Evolution of the Order Primates can be traced back to the early Tertiary period about 65 million years ago. The primates are considered to have developed as an offshoot of a group of small, nocturnal, insectivorous mammals known as tree shrews. A tremendous amount of work has been done on primate evolution, anatomy and physiology, contributing greatly to the overall understanding of primate origins, taxonomy, and phylogeny. Many factors, such as climate, environment, competition, diet, and geographical/behavioral isolation factors, have played significant roles in the origins, biological and physiological evolution of primates. Such factors also account for adaptation,s which can be traced in the fossil record, the loss or retention of ancestral (primitive) traits in some primate species, and in particular for speciation events which led to the divergences of lineages from the earliest protoprimates to anatomically modern humans.

Special thanks and appreciation to the following individuals and departments who assisted in guiding me through the original paper as well as the challenging task of creating this work:

Dr. Michael F. Gibbons, Jr of the University of Massachusetts at Boston for his advice, guidance exceptional sense of humor, and phenomenal patience.

Ellen Royalty , the most creative, perspicacious and patient reference librarian..

Brian Butler, Dave Ford and the Northern Illinois University graduate computer lab for so much time in guiding and assisting with the layout, software, and final printing.

And as always….. my "boys" Jasper and Junior.

"The more complex the mind, the greater the need for the simplicity of play." –Captain Kirk, Shore Leave

Table of Contents

Introduction

The history of primates is primarily centralized within the Tertiary period, beginning at about 65 million years ago when mammalian species began to emerge and diversify. The suborder (or infraorder) Plesiadapiformes is the first group of primate-like mammals known in the fossil record. Plesiadapiformes first appear in the early Paleocene, some 65 million years ago in sediments of the Western Interior of North America. The relationship of Paleocene plesiadapiforms to Eocene primates of modern aspect and relationships among various families, genera, and species of plesiadapiforms are uncertain. In particular, the relationship of Microsyopoidea to plesiadapiforms has been questioned. However, it still stands that plesiadapiformes are the earliest group to emerge from mammals and begin to diverge into other primate lineages, (Herskovitz, 1977; Fleagle et al 1988; Groves, 1989).

The Tertiary period begins with the Paleocene epoch at 65 million years ago and ends at the Pleistocene-Holocene boundary at about 0.010 million years ago,(Berggren et al, 1995). The primates are a diverse eutherian (placental) group with an extensive life history throughout geological time. Many major mammal groups evolved in relative isolation on separate continental landmasses following the separations of the various continents at various points in time. As continents continued to move, as ocean levels shifted or as glaciers formed ice-type land bridges, such formations as isthmus or island (land) chain bridges allowed for some interchange, resulting in extinction of many groups and the divergence of new ones, (Gibbons 1972,1981;Rosen, 1974;Conroy, 1990;Rowe, 1996).

Transitional primate-like creatures, the shrews, first appeared by the end of the Mesozoic Era (65 million years ago) at the K/T boundary. At that time, the world was very different from today. The continents were in other locations and they had somewhat different shapes, there were also fewer of them at that time. North America was still connected to Europe but not to South America. India was not yet part of Asia but heading towards it. Australia was close to Antarctica. Most land masses had tropical or subtropical climates. The flora and fauna at the end of the Mesozoic Era would have seemed alien since most of the plants and animals that are familiar to us had not yet evolved. Large reptiles were beginning to be replaced by mammals as the dominant large land animals. Among the mammals, there were a few archaic egg-layers (monotremes) like the ancestors of the platypus and echidna. The oldest fossil monotremes come from the Lightning Ridge opal fields of New South Wales, Australia. An opalized lower jaw fragment of Steropodon galmani was found to be more than 100 million years old (middle Albian, Cretaceous), (Archer et al, 1992). The first marsupials date from about 100 million years ago (Cretaceous), but they are almost certainly a bit older than that, possibly dating back as far as the Jurassic (160 million years ago). Whatever the case, there was diversification underway by the Cretaceous which set the scene for an explosive radiation in the Tertiary, following the demise of the dinosaurs. Telling marsupial mammals from the placentals in the fossil record is difficult. There are lots of soft tissue and physiological differences but these do not fossilize especially well. There are dental and jaw differences which are useful and a specialized epipubic bone which is sometimes found. Though there were larger numbers of pouched opossum-like mammals (marsupials) in existence, the first placental mammals did not appear until around 70 million years ago, (Rosen, 1974; Gibbons, 1972,1981;Quirk et al, 1983; Szalay et al, 1987; Walker et al, 1989; Griffiths et al, 1991;Pasqual et al, 1992). Those early placental mammals mainly consisted of the transitional primate-like creatures (shrews) and their insectivore ancestors. At thi spoint, the great proliferation of flowering plants had not even taken place as yet,(Archer et al, 1985; Obradovich, 1993; Graddstein et al, 1995).

The first primate-like mammals, or proto-primates, were roughly similar to squirrels and tree shrews in size and appearance. The existing, fragmentary fossil evidence (mostly from North Africa) suggests that they were adapted to an arboreal way of life in warm, moist climates. They probably were equipped with relatively good eyesight (early stereoscopic vision development) as well as hands and feet with pads and claws for climbing, (Birdsell, 1972;Gibbons, 1972; Simons, 1972; Fleagle, 1999). Morphological and functional studies of dental, cranial, and postcranial remains of plesiadapiforms suggest retention of plesiadapiform traits within the order Primates. Four distinct superfamilies are recognized within Plesiadapiformes. Microsyopoidea are characterized by the retention of generalized primitive (relative to Purgatorius) dental and cranial adaptations. Microsyopoids are represented by two families: Paleocene Palaechthonidae and Eocene Microsyopidae. Available evidence indicates that microsyopids are most closely related to palaechthonids and can best be viewed as having descended from that Paleocene family, not from the families Mixodectidae or Leptictidae as previously suggested, (Rosen,

1974;Szalay et al, 1987, Walker et al, 1989;)

The other three superfamilies of Plesiadapiformes are Plesiadapoidea, Mixodectoidea, and Apatemyoidea. Of these, plesiadapoids are characterized by specialized dental and cranial adaptations. Mixodectoids and apatemyoids are distinctly more primitive than either of the other two superfamilies, remaining more insectivore-like. Two families, Microsyopidae and Paromomyidae, survived well into the Eocene, while other plesiadapiform families disappeared by the earliest Eocene. Dental characteristics suggest that these two families specialized on diets different from those of adapids and omomyids and thus avoided direct competition with primates of modern aspect. Geographic distributions and climatic reconstructions indicate that warming temperatures into and through the Eocene contributed to the extinction of most plesiadapiforms, (Simpson, 1953; Rosen, 1974; Eldredge, 1980; Szalay, 1981; Kimball, 1993; Amundson, 1996; Fleagle, 1999).

The first true primates were early prosimians that evolved by the middle of the Paleocene Epoch. Their bones have been found in 60,000,000 year old geological deposits in North Africa. However, they are physically different from today's primate species. Prosimians were still somewhat squirrel-like in size and appearance, but apparently they had grasping hands and feet that were increasingly more efficient in manipulating objects and climbing trees. It is likely that they were also developing effective stereoscopic vision. The beginning of the Eocene Epoch coincides with the appearance of primate species that somewhat resemble modern prosimians such as lemurs, lorises, and possibly tarsiers. This was the epoch of maximum prosimian adaptive radiation. There were at least 60 genera of them in two families (Adapidae and Omomyidae). This is nearly four times greater than the prosimian diversity which exists today. Eocene prosimians lived in North America, Europe, Africa, and Asia. It was during this epoch that they reached the island of Madagascar. The great diversity of Eocene prosimians was probably a consequence of the fact that they did not have competition from monkeys and apes, as such lineages had not yet begun to diverge. Major evolutionary changes were beginning in some of the Eocene prosimians that foreshadow species yet to come. Their brains and eyes were becoming larger, while their snouts were getting smaller. At the base of a skull lies the foramen magnum. The position of this cranial opening, which allows the passage of the spinal column into the cranium, is a strong indicator of the angle of the spinal column to the head and subsequently whether the body is habitually horizontal or vertical . During the Eocene, the foramen magnum in some primate species was beginning to move from the back of the skull towards the center. This suggests that they were beginning to hold their bodies erect while hopping and sitting, like modern lemurs, galagos, and tarsiers. By the end of the Eocene Epoch, many of the prosimian species had become extinct. This may be connected with the appearance of the first monkeys during the transition to the next geologic epoch, the Oligocene (35.4 million years ago), (Simpson, 1953; Rosen, 1974; Groves, 1977; Eldredge, 1980; Szalay, 1981; Bearder et al, 1987; Kimball, 1993; Amundson, 1996; Anderson, 1998; Fleagle, 1999).

The Oligocene Epoch was largely a gap in the primate fossil record in most parts of the world. This is especially true for prosimian fossils. Most of what is known about them came from the Fayum deposits in Western Egypt. While this area is a desert today, 36-31 million years ago (during the early and mid Oligocene) it was a tropical rainforest. Other Oligocene deposits containing some fossil primate bones have been found in North and West Africa, the southern Arabian Peninsula, China, Southeast Asia, as well as North and South America. Monkeys evolved from prosimians sometime during the early part of the Oligocene. They were the first species of the suborder Anthropoidea. Two genera of these early monkeys have been identified, Apidium and Aegyptopithecus. The former was about the size of a large squirrel (2-3 pounds), while the latter was the size of a large domestic cat (13-20 pounds). Both were probably fruit and seed eating forest tree-dwellers. Compared to the prosimians, these early monkeys had fewer teeth, less pronounced muzzles, larger brains, and increasingly more forward-looking eyes. Due to the comparative scarcity of Oligocene Epoch prosimians, it is generally believed that the monkeys out-competed and replaced them in most environments at that time. Supporting this hypothesis is the fact that modern prosimians either live in locations where monkeys and apes are absent or they are normally active only at nighttime when most of the larger, primates are sleeping, (Rosen, 1974;Szalay et al, 1987, Walker et al, 1989).

The Oligocene was an epoch of major geological change with resulting regional climate shifts that likely affected the direction of evolution and altered fossil preservation conditions. It was during that time period that North America and Europe drifted apart and became distinct continents. The Great Rift Valley system of East Africa also was formed along a 1200 mile long volcanically active fault zone between large tectonic plates. Approximately 55,000,000 years ago, India finally came into contact with Asia and began forcing up the Himalayan chain of mountains and the Tibetan Plateau beyond. Progressively, growth of this immense barrier altered continental weather patterns by blocking the summer monsoonal rains. These and other major geological events during the Oligocene triggered global climatic changes. There began a cooling trend, especially in the Northern Hemisphere. A result was the general disappearance of primates from these northern areas. By the middle of the Miocene the continued movement of tectonic plates caused Africa and Asia to reconnect. New mountain chains were forced up and major climatic changes occurred. Much of the East African and South Asian tropical forests began to be replaced by sparse dry woodlands and grasslands. As a result, there were new selective pressures affecting primate evolution, (Herskovitz, 1977; Fleagle et al, 1988, 1999; Conroy, 1992).

Primate fossils are most common from the Miocene. Apes diverged from monkeys early on during this time. Fossil monkeys and prosimians are comparatively rare from the Miocene, but apes are common. Apparently, apes at that time occupied some ecological niches that would later be filled by monkeys. Among the Miocene primates most likely were the ancestors of the modern species of apes and humans. The group of apes that included a more direct lineage leading to anatomically modern humans were apparently in the process of adapting to life on the edges of the expanding savannas in East and South Africa. By late Miocene times, the line leading to humans probably diverged from that of the apes. It is not yet possible to say which of several genera of Miocene apes led to humans or to specific living ape species, (Bown , 1976, et al , 1984 and 1991; Fleagle et al, 1988, et al, 1986; Conroy, 1992).

While there are many physical traits that characterize most living primates, there are very few traits which characterize all primates to the exclusion of all other mammals. Overall, primates can be defined from a purely descriptive sense: Unguiculate, (possessing claws or nails), claviculate (possessing a clavicle), placental mammals, with orbits encircled by bone; three kinds of teeth, at least at one time of life; brain always with a posterior lobe and calcarine fissure; the innermost digit of at least one pair of extremities opposable; hallux with a flat nail or none; a well developed caecum; penis pendulous; testes scrotal; always two pectoral mammae, (Mivart 1873, Le Gross Clark, 1959, Vaughn, 1986). There are also trends in the primate lineage towards shorter snouts, convergence of axes of vision, enlargement of brain, lessening of olfactory ability and prolongation of postnatal growth period (Conroy, 1990). Today there are approximately 233 living species today which have been placed into 13 families. The smallest of the primates is the pygmy lemur, which weighs about 30 grams, the largest is the gorilla, weighing up to 175

kilograms, all also displaying a wide variation of life histories. The fossil record clearly indicates

that distinctively primate, and subsequently human, traits appeared neither recently nor all at

once. Rather, they evolved piecemeal over a period of roughly 5 million years, with the earliest

record of the primates appearance coming in the beginning of the Cenozoic. In terms of primate

evolution overall, and the concept of species for the purposes of taxonomic placement, it is

important to bear in mind that many species, primates being no exception, existed as ring species

at times. Meaning, there were populations that were larger at times during good environmental

conditions and all came together and could breed. During harsh and scarce environmental times,

populations would break apart and small pockets would exist in whatever small niches they

could. This led to some degree of differentiation, some isolation, and therefore some speciation,

while simultaneously creating enough variation within one species yet not creating an entirely

new one, (Hill, 1972; Rosen 1974;Fleagle et al, 1978;Szalay et al, 1987, Walker et al, 1989).

Evolution and Origin of the Primates

Models of the Evolutionary Process

Punctuated equilibrium is a model of speciation, which discounts selection. Speciation is

initiated by a radical mutation, which is said to affect an organism's early development. If the

so-called "mutants" can survive and reproduce long enough to adapt to their environment, then it

is believed that they are likely to be reproductively isolated from the parent species, and thought

to eventually become able to replace the parent species. The result is that punctuated equilibrium

states there is a pattern observable in the fossil record which demonstrates and supports the

sudden appearances of new species, along with the idea that each species changes very little (an extended stasis period) until another radical mutation event occurs, (Gould and Eldredge, 1972).

Phyletic gradualism proposes that beneficial mutations occur once in a while and spread through the population, gradually increasing the variations between it and the population of origin. It further states that the environment is also gradually changing and the population accumulates mutations, which aid in dealing with such events. Therefore, a fossil pattern is created which is similar to that of the one described by punctuated equilibrium (still demonstrating an extended stasis period for a species). According to this model, however, divergent selection and drift quickly act to morphologically and genetically isolate a population from the population of origin, and most evidence collected via lab and nature studies tend to support a more gradual evolutionary pattern, (Moller et al, 1993; Rice et al, 1993). There should be no reason, however, why both selective and non-selective forms of speciation should not occur.

Overall, Eldredge and Gould (1972) claim that changes within a species are neither insignificant nor so great that they warrant the definition of a whole new species. Taxonomy, they claim, has no real application to evolutionary pattern. The reason that few fossil types are found that are transitional between related species, they argue, is that all the changes occur within one or a few generations. Their perception of the taxonomic hierarchy as discontinuous is the basis for their model, as it distinguishes between populations and species. The difference then being that species are separated by full or partial reproductive isolation. Selective changes, they agree, do occur within a species, which results in adaptation, but the changes are not governed by selection (selection merely determines which species survive).

Phyletic gradualism relies upon selection and drift, continuous processes, which include genetic changes that then result in the development of subspecies as well as entirely new species. It predicts that the genes which are being selected upon and which cause such differentiation (divergences) are mostly what are known as additive genes which directly affect the phenotype. Such changes are small, but are numerous. The result is that change will be gradual over the time scale of many generations (geological time) and intermediate fossils will be found in the fossil record, (Moller et al, 1993; Rice et al, 1993; Weiner, 1995).

Charlesworth (1990) described a combined mechanism for speciation, which was a weak attempt to combine the two models. His error was that he also failed to account for small populations or organisms existing and perhaps being left out of the fossil record altogether, or the fact that minor changes occurred both within the environment and the populations, oftentimes in response to one another. Therefore, this also led to a similarly distorted view of evolution's mechanisms. Species tend to become separated from one another for a variety of environmental and adaptational reasons, and each is held together then via gene flow, selection, and their developmental processes. There is a constant fluctuation within a species, meaning that there is constant action by selection pressures in response to continuous changes in the environment. So, evolution in actuality is a continuous process, as these fluctuations are minor. Allele frequencies for various traits are under constant selection pressures to either increase or decrease in response to the environment. Minor mutations occur which are acted on in a similar manner as well, which can be either beneficial or harmful to the population. Also, major geological events can occur which can leave a small population or a smaller organism completely out of the fossil record, particularly if either enough offspring were not produced or preservation of remains was poor. In which case, only macromutations would become more apparent in the fossil record, supporting the punctualist's position, yet upon close examination of marine microorganisms a more gradual pattern can be detected. Thus leading to the belief that perhaps a combination of models is a more accurate representation of the true mechanisms of evolution, (Simpson, 1953;

Hoffman, 1989; Charlesworth, 1990). Neither punctuated equilibrium nor phyletic gradualism is well supported, and it may be impossible without combining the two to ever be able to estimate rates of speciation and discover what actually may occur between speciation events.

Arborealism

Primates are primarily arboreal mammals. The earlieset primates evolved most likely from an insectivorous mammalduringthe arly part of the Paleocene.The fossil record makes it difficult to determine precise realtionships between primates and other orders of mammls, however, it is the same fossil record which makes use of comparative anatomy , along with paleontological and molecular studies, that can provide information as to the degree of relatedness and the shared evolutionary histories between primates and non-primate mammals, (Fleagle 1988; Begun et al, 1997). The few primates that live terrestrially still display an arboreal ancestry. Primate ancestors were generalized arboreal mammals with longer muzzles and laterally placed eye orbits. They were structured much like the tree shrews of Southeast Asia,with short limbs and bushy tails as well. Since primates have not been the only mammals to have lived in the trees, yet display such adaptations, it is likely thattheir speciizations are not merely adaptations to arboreality, but to a specific means for being able to conduct life in the trees overall. Cartmill, (1974), has shown that a likely explanation for the specialized features of primates has been due to what is known as visual predation, which has been correlated with insect hunting while living in the trees. Therefore, to a tree-swelling predatory species certain specializations would be advantageous. Such adaptations would include: in terms of locomotion, the ability to grab and hold branches in addition to prey, thus accounting for the development of the prehensile tail, and to provide a strong hold on branches, the big toe is separated from the other toes in all species except humans, and the thumb is always separated from the fingers, although it is fully opposable only in apes and in some Old World monkeys, and the arm and wrist bones are not fused, which increases dexterity. In those with prehensile tails, the tail acts as a fifth limb for the purposes of grasping as well as balance during locomotion.   Since primates climb with their hands and feet as opposed to claws such as the case with mammals like the squirrel, the thumb and big toe close in opposition to the other digits. Primates also have flexible shoulders for the accomodation of locomotive behaviors such as swinging, hanging, or brachiating. In trees the predation hypothesis prevails since the sense of sight is the most depended on in arboreal habitats. Therefore, the eye orbits evolved with a trend toward more forward directed eyes (orbital convergence). This provides overlapping three dimensional vision. Unlike other mammals, this visual adaptation led to protection for the eyes in the development of the postorbital ring which surrounds the eye. The mammalian longer snout then gave way to the smaller primate ones due to an increased dependence on sight rather than smell. Primates also developed specialized structures, Meissner's corpuscles, in order to reduce slippage and increase sensitivity to arboreal support. Since these arboreal environments encompass their own varying degrees of environmental diversity, so do thier primate and non-primate mammalian inhabitants. It was over the course of varying geological events over time which drove the development of such adaptive diversity, (Gregory, 1951;Biegert, 1961;Hennig, 1965;Cartmill, 1974; Fleagle et al, 1980, 1988; Conroy, 1990). The Cenozoic era (65mya - 10,000ya), (Park , 1992), is considered to be the "age of mammals" due to the fact that prior to this time none existed in the fossil record.   It was just prior to this, at the tail end of the Mezozoic, that therapsids began to diverge and give rise to early mammals. During the Cenozoic's first period, the Tertiary (65mya -5.2mya), (Park, 1992), proto-prosimiams , true primates, and finally hominoids began to appear. During it's second, the Quarternary (1.6mya-10,000ya), (Fleagle, 1988; Park, 1992), hominids, more modern mammals appeared. These protoprosimians, or pleisiadapiformes , emerged during the Paleocene epoch , just following the massive extinction of the dinosaurs, with the true prosimians showing up by the Eocene, about 56.5mya. By the Oligocene, (35.4mya), the anthropoids begin to emerge. Anthropoids being the lineages which later diversified and led to apes, gorillas and humans, (Fleagle, 1988; Park, 1992).

What accounts for both such radical and subtle changes in the climates over time , a factor in extinctions as well as in diversification of species, is the dynamics of the earth's internal and external sturctures. There are what's know as plates which comprise both the terrestrial continents and oceanic floors. It is the constant motion of these plates which brings about environmental   changes and affects migration as well.

Continental Drift

The separation of the continents of the Paleozoic (570-438 million years ago), (Laing, 1991), after having drifted apart through the fragmentation of the supercontinent Rodinia, (Windley, 1984), drifted together again During the Paleozoic, colliding to form the supercontinent, Pangea During the Devonian (408-360 million years ago), (Laing, 1991), and Carboniferous (360-286 million years ago),(Laing, 1991). More specifically, Pangea was assembled by the collisions of three main blocks, Gondwananland, Laurussia, and Siberia, During the Permo-Carbiniferous time of about 350-260 million years ago, (Irving, 1977;Windley, 1984;Laing, 1991). Various smaller blocks of land also contributed to its overall formation. During the Ordovician period of around 500 million years ago,(Irving , 1977), the continents of Laurasia and Siberia collided. The next impact occurred during the lower Devonian. This first collision is known as the Acadian orogney (as it encompassed what is now the Maritime Provinces of Canada) and it continued throughout the Devonian and into the Mississippian (360-320 million years ago),(Laing, 1991). Most of the ocean that had once separated them had vanished as mountain ranges formed. Simultaneously, Gondwanaland was moving across the South Pole and northward towards what is now the South Atlantic. Gondwanaland included all of Africa, South American, Antarctica, Australia and New Guinea, (Du Toit, 1937; Creer, 1970; Burrett, 1974). The second collision, the Appalachian orogney, involved Gondwanaland was moving north, and Laurassia was moving as well, and was in its path, (Hume, 1948; Morel et al, 1948). Laurassia's contact with Gondwanaland is thought to have occurred at some point During the Late Mississippian in the area where Oklahoma now sits, and also in the Early Pennsylvanian (320-286 million years ago), (Laing, 1991) in the present area of the Appalachians. Other timings throughout such areas as Europe are not well known, but the collision ultimately resulted in mountain ranges extending from most of North America through France and into Eastern Europe, (Churkin, 1973). While this collision was occurring, smaller continents such as Angaria and Baltica were uplifiting the Urals and the mountains of Nova Zemlya, (Du Toit, 1937; Churkin, 1973; Bridges, 1990). All of these collisions combined, formed the one continent hypothesized by Wegener to represent Pangea. Wegener compiled evidence which included such things as ancient glaciations, the distribution of fossils, the grooves and cuts found along coastal edges of the then existing continents, along with the fact that preservation was key to fossil evidence. It was Rodinia that is believed to have separated and became Lauassia and Gondwanaland. Laurassia is the proDuct of North America and Eurasia coming together,

(Tarling et al, 1975; Raymo, 1983; Windley, 1984). Gonwanaland included all of Africa, India, South America, Antarctica, Australia, New Guinea, and New Zealand, (Tarling et al, 1975: Raymo, 1983; Windley, 1984). Evidence that Pangea existed can be found when discussing land animals, vegetation, mountains and climate. Early observations uncovered two very different floras which flourished about 350-220 million years ago, one in Laurassia and one in Gonwanaland. The Laurassian flora was dominated by large scale trees related to ground pines, found to be more tropical. Examination of annual growth rings revealed that the equator was at one time in Laurassia during the late Paleozoic. In contrast, the flora of Gonwanaland was dominated by large seed fern trees which had stem wood displaying well developed growth rings indicating a more temperate climate. Given the unequal distribution of heating throughout the continental and oceanic crusts of the earth, this could be explained as different regional climatic environments locally responding to the internal planetary forces which in tern also affected the weather conditions regionally. Since vast amounts of land comprised this one continent, it is reasonable to believe that local conditions would vary resulting in such differences in flora and perhaps even to some extent in the land animals found throughout it. The other possible secondary explanation would lie in the abilities of the seeds to be transported throughout the continents. This would be dependent upon weather patterns and the types of land animals in existence at the time. Seed dispersals in localized areas would be more possible than dispersal over long ranges. Two reptiles also provide more convincing evidence for the existence of Pangea. *Mesosaurus* lived During the Permian (286-245 million years ago), (Laing, 1991), was unsuited for marine environments Due to its body conformation, and perhaps even most likely unsuited for lengthy transcontinental treks, yet evidence suggests that over time it was able to accomplish them. Fossils of *Mesosaurus* have been found in both Argentina and South Africa.

This is also true of *Lystrosaurus* , a wholly terrestrial animal with a similar build, and whose fossils appear in disjunct locations separated by thousands of miles of ocean. One of the greatest challenges of the continental drift hypothesis was to see if *Lystrosaurus* fossils could be found on Antarctica, as it had already been found on other Gonwanaland continents. In December of 1969, such fossils were unearthed, despite the fact that the search was unintentional. They were found 650km from the South Pole in the Transantarctic Mountains, (Du Toit, 1937; Churkin, 1973; Windley, 1984; Bridges, 1990;de Blij,et al,1996). Other evidence points to the formation of mountain ranges mentioned earlier, ranges which are now separated by large bodies of water, such as the Appalachians which are found in North America as well as the British Isle and Scandanavia. The mountains have been found to be of the same age and structure, and if fit together would form a continuous belt. In terms of climatic evidence, the Southern Hemisphere was once by Antarctica and the Norhtern by the equator. There is evidence that ice sheets once covered the Southern Hemisphere, which is now an equatorial region, and the large tropical swamps of the Northern only exists as faunal remains, (Tarling et al, 1975;Raymo, 1983; Windley, 1984).

The Pangean supercontinent lead to many changes in the shape of the land, glaciation patterns and climate, which in turn altered sea levels and increased ocean salinity. The formation of Pangea led to the initiation of some extreme environments, and along with volcanic activities related to the impact this formation is often seen as the cause of many Permian mass extinctions. During Pangea's formation, there was massive cooling and an increase in glaciation. There was also a drop in sea levels and a loss of shallow bodies of warm water, as well as lost continental shelf habitats. Fossil, faunal and geological evidence supports not only such a formation, but the interrelatedness of a mass extinction, which also left behind much of the support evidence of the formation. It is this example of continental drift that explains major climatic changes as well as accounting for a variety of seemingly Old World or Prosimian fossils found in various parts of North America. This supports continental drift as a major factor of climatic change leading to alterations in migratory patterns of mammals as well as early primates, (Raymo, 1983; Windley, 1984; Fleagle 1988, 1999).

Rooneyia

The Rooneyia omomyid skull is a fossil tarisform primate of Europe and North America (Texas)of the Oligocene, (Wilson, 1966),(though some literature lists it as Eocene). The first unequivocal primates occur about 50 mya. There are two main groups identified: Adapiformes which are usually considered to be ancestral to modern Strepsirhines; Tarsiiformes which are (mostly) considered to be early Haplorhines. Omomyids are the best examples of early Tarsiiformes. For example Rooneyia or Necrolemur. These early Tarsiiformes have some features to associate them with later anthropoids, including short face, big eyes, narrow gap between eyes, large brain, and tubular ectotympanic bone. The finding of Rooneyia's skull provided evidence that the continents were most likely one when early forms began to migrate on a global scale, (Bown, 1976;Fleagle, 1988;Rose *et al,* 1991; Gunnell, 1995).

Primate Origins

Early primates evolved from insectivorous mammal ancestry at some point during the late Cretaceous to early Paleocene (about 73-65 million years ago) (Fleagle, 1999). Through the use of comparative morphological studies based upon fossil evidence, molecular data, and morphological data from living primates, inferences can be made about the relationships between primates and other mammals. The superorder Archonta includes Primates, Scandentia (tree shrews), Dermoptera (flying lemurs) and Chiroptera (bats). These suboders are felt to be the most closely related with respect to primates, (Wible et al, 1987; MacPhee,1993; Buckley, 1997;Fleagle 1999). The Plesiadapiforms (primate-like mammals) have been thought to have played a role in primate origins, yet have now been removed from Primates to their own order, Plesiadapiformes. They, as are the other orders, very close to the divergence times when primates began to emerge. However, they may still be the best known mammals of the early Cenozoic and are still an important part of primate origins. Much of what is known of primates origins stems from dental comparisons between both fossil and extant species of primates and other mammals, and most group early primates into adapids (lemur-like) and omomyids (tarsier-like). Plesiadapiformes gave rise, however, to both adapids and omomyids, and therefore are considered early primate forms or a type of primate-like mammal. Despite phylogenetic problems in the tracing of primate origins, it is generally felt that ecological factors were the most important driving force behind their emergence, (Szalay,1972; Cartmill, 1974;Sussman et al, 1978).

Very late in the Cretaceous period the first placental mammals and early mammal-like primates appear, at about 73 million years ago (Fleagle 1999). The Paleocene contains the best documented evidence of these first primates, the Plesiadapiformes. Ecologically, this was a time of warmer climates and the development of both tropical and semi-tropical forests. The majority of the mammals at the time were insectivorous in nature, and at their peak during the Eocene, were globally widespread, more so than the extant species today, (Szalay, 1972;Gingerich, 1976; Sussman et al, 1978). From these insectivores, some feel early primates emerged. These early primates adapted to the now more arboreal environments in several ways. Their orbits converged more forward and towards the center of their skulls leading to an increased reliance on visual senses. This in turn led to the facial and jaw reduction associated with tooth reduction in primates. Also, grasping hands and feet with nails rather than claws developed, (Cartmill,1972, 1974; Conroy, 1990).

The fossil mammals that are most likely the common ancestors of the prosimians are the plesiadapines, which originated about the middle of the Paleocene (roughly 80 million years ago) (MacPhee 1993;Fleagle 1999). Their cranial capacity was small, and they possessed elongated muzzles with a short-tailed, elongated body comparable in size to a squirrel. They had specialized anterior teeth which were rodent-like. Their incisors were elongated and chisel shaped, with molars that were wider than their length with pointed cusps for shearing. Further evidence indicates that they also possessed claws rather than nails on their digits. Today it is not felt that these primate-like mammals led to the prosimians of today, but that they led to a similar intermediate of more direct ancestry, (Van Valen et al, 1965; Wible et al, 1987; MacPhee,1993; Fleagle 1999). One of the earliest plesiadapids was *Purgatorius,* from the earliest part of the Paleocene. It has a dental formula of 3-1-4-3 with molars that are more like those of *Ptilocercus* (a member of the Tupaiidae, tree shrews). *Purgatorius* has been thought to have diverged from this group of insectivores, leading to primate development. There are other intermediate forms linking *Purgatorius* with both the insectivores and other plesiadapiformes. Such identifying common features for this basis include relatively lower molar trigonids, broad second lower molars, and elongated third molars, (Clark, 1934; Van Valen et al, 1965;Beard, 1990; Buckley, 1997).

The Eocene brought about the expansion of rodents into the niches of early primates and the disappearance of land bridges between continents. It became more evident that early primates of this time did begin to adapt to a more arboreal life, altering them morphologically as previously discussed. Some important adaptations to emerge as a result included larger brains, orbits shifting more forward in the skull, and a more forwardly placed foramen magnum (demonstrating a shift to a more upright position), (Van Valen et al, 1965; Cartmill, 1974; Gingerich, 1986;Wible et al, 1987; MacPhee,1993; Fleagle 1999). One fossil primate group to have been derived from the plesiadapiformes were the Adapines (Adapidae). Like the plesiadapines, they were also globally spread and were represented by a number of lemur-like primates, the best known being *Adapis*, whose body proportions and size were similar to those of the extant lemurs (30g - 2kg)(Sussman et al, 1978). *Adapis* lacked the prosimian toothcomb (procumbent lower incisors), yet possessed the first known opposable digits, and were ancestral to the extant lemurs and lorises, (Gingerich, 1986;Wible et al, 1987; MacPhee,1993; Fleagle 1999).

   Fossil tarsiers from the Eocene found in both Europe and North America showed considerable amount of variation in that they tended to have larger forebrains and shorter faces than the lemurs, and evolved into vertical climbers and leapers. Though it is not known as yet if these Eocene tarsiers evolved from the lemuroid mammals or directly from tree shrews, (Van Valen et al, 1965;Sussman et al 1978; Gingerich, 1986). The Eocene was likely the time when the ancestors of both New World and Old World monkeys were undergoing parallel evolution. These ancestors are collectively known as the Omomyids, and inhabited North America, Europe and part of Asia. They have been thought to also be ancestral to the anthropoid apes (Fleagle 1999). The earliest of the apes were a fossil species from Burma, *Amphipithecus,* who had three premolar teeth yet displayed some advanced morphology, and the less well known *Pondaungia*, also from Burma, (Bown, 1976; Bown *et al*, 1987; Wible *et al*, 1987; MacPhee, 1993; Fleagle 1999).

Primates are usually grouped with Plesiadapiformes, Scandentia, Dermoptera and Chiroptera in the superorder Archonta. This grouping was based upon many aspects of postcranial anatomy and a few features of the skull and dentition. The phylogenetics have as yet to be ironed out, since some feel the basis of relatedness lies in skeletal features, while others feel the basis lies in behavioral traits (such as dietary and habitat choices). However, fossil record of early primates still leans toward the indication that primates were derived from placental mammals which arose in the Cretaceous, and which gave rise to the orders discussed, particularly the tree shrews and their subfamilies, (Bown, 1976; Fleagle, 1988; Rose, 1995).

Tree Shrews

The eighteen species of tree shrews, Family Tupaiidae, are scattered throughout southeastern Asia and extending into Malasia and Indonesia. Their body size and habits are much like squirrels. Little is known of their behavior or reproduction. It was long thought the group was closely related to Primates and Insectivora, perhaps forming a link between them, and various literature has at various times have placed the tree shrews as a family within each of these orders. They have also been linked by various others to bats, elephant shrews, and dermopterans (flying lemurs).It is also quite possible that tree shrews are an example of a mammal in transition which may have not only given rise to a number of taxonomic orders, but retained many primitive traits which has led to such confusion over its own taxonomic placement. Both the dentition and ovarian bursa features of these animals act as indications that this is most likely the case, and that the tree shrew is indeed a primitive primate, (Clark, 1926, 1934;1959; Gibbons, 1981; Fleagle, 1988; Wibble *et al,* 1994).

The fossil record of tree shrews extends only into the Pliocene, and the modern species are closely similar to earliest fossils. Early on, this scanty record presented real problems with linkages, as noted above. The best position at present is to consider this a very conservative group in its evolution, (meaning they have retained primitive traits). Arboreal species generally present a problem with preservation of fossils, so early relationships based on anatomical evidence from modern species may remain enigmatic. It is possible that such a group gave rise to at least three major orders, Primates, Insectivora and Tupaiid, (Clark, 1926, 1934;1959; Gibbons, 1981; Fleagle, 1988; Wibble *et al,* 1994).

The Tupaiiformes are not in actuality shrews or arboreal mammals. They live in bushes and lower branches of trees in tropical rain forests. One species, *Tupaia glis,* may be a ground dweller in certain regions. The biggest problem surrounding these creatures however, is what has already been mentioned above, are they primates, insevtivores, or something in between, (Huxley , 1872; Clark, 1926, 1934;1959; Gibbons, 1981; Fleagle, 1988; Wibble *et al,* 1994).

There is a definitive tree shrew pattern, and they have a wide distribution range throughout Southeast Aasia and some parts of Asia, India, Ceylon, Vietnam, Cambodia, and the Phillipines. They live in primarily the tropical rainforests which border mountainous terrains. Generally, the smaller species of tree shrew, the closer it lives to the ground, and at best can only be considered partly arboreal. The tree shrews at first have a rodent-like or squirrel-like appearance. In fact, the Malay word "tupai" means squirrel, (Dolhinow, 1972). They have long bodies and tails with dull coats, and little sexual dimorphism is evident. The muzzle is elongated and expands into an extensive rhinarium (or the naked area like a wet nose). The upper lip is attached to the maxilla and therefore is said to be teethed. This limits facial expression. Unlike true primates, all digits are clawed. They can not pick up objects or climb. The hands are not truly prehensile, although a convergent grip of both hands is utilized., and there can be at times a tendency towards the independent movement of digits. The thumb is divergent from the other digits, but is not opposable to them. The pen-tailed treeshrew is the one exception, as it is known to be somewhat aberrant in that its thumb and big toe are opposable, thus making its hands and feet prehensile, (Huxley , 1872; Clark, 1926, 1934;1959; Dolhinow, 1972).

Tree shrews are quadrupeds and their extremities are rather short with respect to their bodies, with the hindlimbs somewhat longer than the forelimbs. They are diurnal, again, with the exception of the pen-tailed shrew. The tree shrews incisors are primate-like in being not chisel shaped, but generalized. The lower incisors tend to be procumbent (directed horizontally) and are used to comb their body fur (ie: tooth comb teeth). Like many prosimians, tree shrews also posses an additional sublingual organ (an extra tongue) that is thought to act as a type of toothbrush for the cleaning of the lower incisors, (Schultz, 1969). They are omnivores, consuming mainly soft foods such as fruits, vegetables, insects, and other various smaller animals. Compared to other mammals of similar size, the tree shrew possesses a relatively large brain. It is a brain which displays some degree of reduction in the olfactory area, and in the visual area some degree of expansion is evident. However, the tree shrew brain is still considered more similar to the brain of the insectivore. The visual expansion though is further seen in the relatively large eyes placed in the "primitive" lateral position, with the exception yet again being the pen-tailed shrew who's eyes are placed somewhat more forwardly. Unlike the generally accepted primates, tree shrews are not thought to have stereoscopic vision, but they do possess color vision ability. Again, unlike most primates, they also tend to have multiple births, with as many as three or four young at a time. Several species have primate -like placenta and fetal membranes, and the retained (primitive) trait of the presence of an ovarian bursa. Multiple pairs of breasts are also found, with some species having as many as three pairs. Another interesting feature many species posses is a throat glad, which has been thought to be for the purposes of territory marking. These traits comprise what is considered the tree shrew pattern, (Huxley , 1872; Clark, 1926, 1934;1959; (Schultz, 1969;Dolhinow, 1972; Gibbons, 1981; Fleagle, 1988; Wibble et al, 1994).

When G.G. Simpson (1945) produced his classification of mammals, he placed tree shrews in the order Primates. Since this time, a continuing debate has been waged as to whether or not this is so. Are tree shrews primates, insectivores, or something in between? Did they perhaps actually give rise to multiple orders, such as Primates, Insectivora, or Tupaiid and Scandentia? It is most likely so given the retention of traits and the bits and pieces of scattered commonalities with such multiple orders. Simpson was not the first however to suggest that tree shrews had primate affinities. In 1872 T. H. Huxley had pointed out certain primate-like features. In 1910 W. Kaudern did as well, as did A. Carlsson in 1922 and Le Gros Clark in 1934. William Gregory of the American Museum of natural History also favored such a placement, (Dolhinow, 1972). It was during the latter half of the twentieth century that the debate over tree shrew taxonomy blossomed, and subsequently rekindled interest in the primates overall.

Some have placed the treeshrews , along with a group known as the elephant shrews, in the separate order , Menotyphla. This has not proven satisfactory because such animals are not all that closely related. The tree shrews are a mixture of primitive mammalian and incipient prosiniam characteristics, and most taxononomies place them in either Primates or Insectivora. It was William Straus Jr. who proposed they be placed in a separate order of their own, Tupaioidea,

(Dolhinow, 1972). It was W.C. Osman Hill (1953) who felt that if tree shrews were to be included with Primates, then the entire order would be left vague and undefinable. Not only anatomically, (as was the case in the retention of the ovarian bursa, the lacking of the os penis and clitoris bones, vision and digit differences), but also behaviorally, the tree shrew pattern is ambiguous. Shortly after giving birth, the mother abandons her young and only returns about once every two days for brief feeding encounters. None of the bonding habits of primates (grooming and nuturing) take place. However, it is in this sea of ambiguity that evidence is found for the divergence of tree shrews from early mammalian ancaestors into several orders, Tupaioidea, Primates, Insectivora, and Scandentia, (Huxley , 1872; Clark, 1926, 1934;1959; Gibbons, 1981,1999; Fleagle, 1988; Wibble *et al,* 1994).

Regardless of the arguments, there tends to be agreement on at least the one point that tree shrews were probably very close anatomically to the extant tree shrews. These first primates (about 70 million years ago) (Dolhinow, 1972) were most likely ground dwelling quadrupeds who became partially arboreal and even at that some to the extent of fully arboreal with expansion of such populations as the rodents who were forcing tree shrews from their terrestrial econiches. It is quite possible that rodetns separately diverged from the same common insectivore ancestor of todays' tree shrew. This makes the tree shrews an excellent living fossil , or rather more precisely, a good structural ancestor for the primates. By this it is meant that such an animal need only posses a total morphological pattern which would have likely been posses by the true ancestor. Therefore, a structural ancestor can be a fossilized animal or an extant species. The tree shrew is an ideal candidate for such a role.

Early Primates

As previously stated, the history of the primates is primarily featured in the Tertiary period beginning about 65 million years ago with the massive radiation of the mammals following the extinction of the dinosaurs, (Fleagle et al 1988; Groves, 1989). The Paleocene (first epoch of the Tertiary, 65-55mya) primates are referred to as Plesiadapiformes. At least one Paleocene family, Plesiadapidae, possess molar teeth and a bony ear region similar to today's extant primates, establishing a base for a lineage. The lineage problem is an interesting one, as the tendency is to attempt to try to directly link extinct primates directly to known living primates. The problem here is that many fossil groups , including those currently under study, are likely to have become extinct without having left direct descendants. That in mind, then, what is known today of primates is only a small sampling of a larger adaptive radiation of animals, and that any fossil group seen may not itself have been directly ancestral to anything now living. There are always many more species that were not ancestors than those which were, and it should be noted that such differences are appreciably subtle since they were all part of the same radiation of closely related animals, (McHenry, 1975; Delson1981; Andrews *et al*, 1987; Beard *et al*, 1991).

It was during the Paleocene when many of the anatomically modern mammals began to develop. These early primates had very few , if any, recognizable features which primates are known for today, hence the tree shrew difficulties previously discusses, as many looked superficially like rodents more than like any known primates. These primates are considered "archaic" or "primitive". They contrast with those who appear more anatomically modern, the euprimates, greatly. Their eye sockets were not completely encircled in bone (postorbital enclosures), their digits were clawed and lacked nails, their faces much more prognathic, and incisors were enlarged and elongated much like those of rodents. It is important to note that these do not constitute synamorphies (shared , derived traits) , but are primitive mammalian traits with the exception of the incisors, which most likely were an adaptive feature of rodents. It is this group, these archaics , which are the base of the primates, (Le Gros Clark, 1934, 1959; McHenry, 1975; Herskovitz, 1977; Delson1981; Andrews *et al*, 1987; Beard *et al*, 1991).

From the Euocene then (55-38 mya), (Park, 1999), that the euprimates first make their appearance. They are the group most phenotypically primate, as they possessed the postorbital bar, nails rather than claws, shorter faces, forward directed orbits, and a grasping big toe. The Eocene primates are known from most of the Northern Hemisphere and fall into two groups, adapids (Adapidae) and omomyids (Omomyidae), (Herskovitz, 1977; Beard *et al,* 1991).

Adapids and extant lemurs and lorises appear to be linked in terms of the anatomical aspects of their ankle and wrist joints. Due to the fact that the links demonstrate a closer tie to the European adapids rather than those found from North America, it appears that the origin and divergence stems from Europe from the ancestor of extant prosimians prior to the formation of a tooth comb.   One of the best known fossils is *Notharctus*, an adapid from North America, whose limbs were indicative of the ability to grasp. It also had ridges on its teeth which are markers of a folivorous diet, along with diurnal eye orbits directed more forward. It had longer hindlimbs thought to be evidence of leaping and quadrupedal running. It further appears to have had a small opening for a maxillary nerve suggesting a reduction in the emphasis on the use of tactile whiskers of prosimians. Some adapids also appear to have had a fusion of the mandibular symphysis, similar to anthropoids. Others have the stapedial artery as the main blood source to the brain , as lemurs today do. On the other hand, some still utilized the promontory artery like the anthropoids. Therefore, adding up to the possibility that the adapids are similar to members of a radiation that ultimately diverged into quite different primates, (Le Gros Clark, 1934, 1959; McHenry, 1975; Herskovitz, 1977; Delson1981; Andrews *et al*, 1987; Beard *et al*, 1991).

Omomyids provide a deeper mystique. While the adapids had the ring shaped tympanic bone as the lemurs and lorises do, the omomyids had a tube shaped tympanic bone, similar to most anthropoids. The *Necrolemur* has long been thought to be an Euocene relative of the tarsier. The basis for this lies in the interpretation of three synamorphic features; elongated ankle bones, fused lower leg bones (tibia and fibula) in some , and enlarged orbitals. Still, when examining the skull base, it almost appears that the differences are so vast that the omomyids are not at all closely related to extant primates. It is therefore crucial to bear in mind the fact that over geological time there have been numerous radiations and adapations to specific environments have occurred as a result. The skull of *Shoshonius (* an omomyid genus) has reinforced a link to tarsiers. The Euocene then of Southeast Asia may have been the home to the stem lineage of anthropoids, as both teeth and jaw fragments of two genera, *Amphipithecus* and *Pondaugnia* have suggested strong anthropoid relations despite the fact that little is known of them still, (Le Gros Clark, 1934, 1959; McHenry, 1975; Herskovitz, 1977; Delson1981; Andrews *et al*, 1987; Beard *et al*, 1991).

Parapithecidae

During the Oligocene, (38-23mya), (Park, 1999), the primary radiations of both the anthropoids and catarrhines took place. The majority of the information pertaining to this time period comes from a single site, the Jeble Qatrani Formation of the Fayum region in Egypt. Studies have shown that 30-40 million years ago, this presently arid region was a swamp area bordered by forests. Plants similar to present day Southeast Asian plant life could be found along with fossil water birds (mainly jacanas and shoefilled storks) , which is suggestive of an environment conducive to the niche needs of modern primates,(Howell, 1977; Kay *et al*, 1980 Simons, *et al*, 1987,1989; Bown *et al*, 1995).

Despite the fact that it is possible the first anthropoids appeared in the late Eocene, they have been identified in the earliest parts of the Oligocene as well. The Parapathecidae comprise several genera, of which *Qatrania* is the earliest and smallest, particularly in comparison to any extand Old World Monkey. Its relative, *Parapithecus,* was about ten times larger, howeverm, which indicates that then , as opposed to now, primates tended to diversify in appearance. *Parapithecus* had lost its lower incisors completely, which is an exception among primates and to some excludes *Parapithecus* from the direct ancestry of anthropoids, ((Howell, 1977; Kay *et al*, 1980 Simons, *et al*, 1987).

The most well known of these parapithecids is *Apidium*, which displays several derived features for anthropoids: a fused left and right side of the mandible, fused left and right frontal bones of the skull, and an eye orbit entirely sealed off from the rear. What is most interesting is its lower leg and how the lower leg bones were pressed tightly together for just about their full length but not fused into a single bone, (Harrison, 1987). This suggests a very limited range of mobility to the ankle joint, and therefore indicates the likelihood that it made its way via leaping through branches. Its smaller eye orbits suggest a diurnal lifestyle, while the thicker tooth enamel suggests hard foods in the diet, (Olson, 1981, Harrison, 1987;Tuttle, 1988).

All the parapithecids had three premolars per jaw quadrant, which makes them likely ancestors of both catarrhines and platyrrhines (loss of a premolar is a derived feature in catarrhines). They do not seem to show derived features that would place them specificaly in the catarrhine lineage though, despite the fact they are obviously found in the Old World. In nearly all ways in which catarrhine primates have a derived anatomical feature and platyrrhines havea primitive one, *Apidium* has the primitive, platyrrhine feature. This is particularly so for the inner ear details, which are often diagnostic of primate taxa. The ischial tuberosities of the pelvis or catarrhines are also absent from the pelvis of *Apidium*. Some dental details such as a new cusp , the hyperconulid , and a particular pattern of wear on the teeth unique to catarrhines have been taken to indicate that parapathecids may have been on the catarrhine lineage after the divergence of the Platyrrhini. The majority of the anatomy however, indicates that these were primitive anthropoids , and that catarrhines and platyrrhines were each other's closest relatives with the parapathecieds representing an early and primitive out-group, (Howell, 1977; Kay *et al*, 1980 Simons, *et al*, 1987,1989; Bown *et al*, 1995).

Catarrhines

Another type of primate coexisted in the Fayum at the same time as the known parapathecids. This kind of primate, known collectively as the Family Propliopithecidae, or propliopithecids, is best represented by *Aegytopithecus zeuxis*, (Kay *et al*, 1981;Simons, 1995; Fleagle *et al* , 1999). *Aegyptopithecus* had lost its anterior premolar, so it had a dental formula of 2.1.2.3, which is considered synapomorphic with other catarrhines, placing this specimen clearly on the catarrhine lineage. Another possibly synapomorphic feature indicating a relationship specifically to the catarrhines involves the arrangement of skull bones at the side of the skull. In most other repsects though, the skull is still considered primitive, along with the limb bones, more in accordance with New World monkeys and the parapathecids. In the ear region, this specimen had a ring bone similar to modern platyrrhines as opposed to the tube which is characteristic of living catarrhines. Also, just above the elbow, *Aegytptopithecus* had a small hole referred to as the entepicondylar foramen, which is primitive and not present in extant catarrhines, (Szalay *et al*, 1979; Fleagle *et al*, 1980,1982, 1999; Simons 1995).

Sexual dimorphism also could be seen in Aegyptopithecus. This implies a type of polygynous social structure analogous to most modern sexually dimorphic Old World primates, (Andrews , 1985). Old World primates, in addition to several skulls, postcranial remains include bones of arm , foot, and tail. Overall, morphology is most consistent with an arboreal, quadrupedal, diurnal animal who's diet consisted of primarily fruits. The Oligocene primate fauna has been lacking though in some of the characcteristic features of catarrhines. There is , for example, no evidence that any of these species was particularly large, despite known variations in size. There is also little evidence thus far of adaptations for leaf eating, seed eating, or terrestrial habits. All features which are common in their descendents, yet Aegyptopithecus is still considered one of he closest relatives to extant catarrhines,(Schwartz *et al*, 1978; Szalay *et al* 1979; Kay *et al*, 1981; Simons, 1995).

Origins of Platyrrhines

During the late Eocene (30-38 million years ago) the higher primates (anthropoids) began to develop from their prosimian ancestors and, with the aid of continental drift, diverged into the platyrrhines (New World monkeys). These platyrrhines of South and Central America are now   a diverse radiation of these higher primates which has evolved in the New World over approximately the past 28-25 million years (beginning during the late Oligocene/early Miocene). However, the fossil record remains quite poor and their phylogeny unresolved. What does exist in the fossil record however is suggestive of an African origin, and in turn, is also supportive of continental drift, leading to the origins of platyrrhines and the eventual dispersment and diversification seen today, (Ciochon et al, 1980; Fleagle, 1999;Houle, 1999).

The somewhat ambiguous ancestry of the platyrrhines derives from a low abundance of fossil material, as well as limitations based upon the knowledge of continental drift. Similar biological diversity and geology suggest that the continent of South America was once a part of Africa. It eventually broke away from the mainland, migrating to it's present position. This took place over 100 million years ago, far earlier than the first primate fossils on either continent on either continent. This has led to two main hypotheses for the ancestry of platyrrhines. One, that they evolved from similar ancestors as the Old World monkeys, a "proto-primate" of sorts which evolved in North America. However, due to the severe lack of fossil evidence, and given the wind and water current data of the time, this hypothesis has been virtually discarded. The second, and more popularly accepted and more plausible hypothesis, is that of an African origin. The splitting of the continents left approximately 600 miles between them, and groups of primates have been believed to have "rafted" across on large clumps of vegetation and pieces of land. Given the data on the wind and water current at the time, this is indeed a possibility. Also, there is fossil evidence supportive of morphological similarities between African anthropods and none of Old World monkeys in South or Central America, suggesting an African origin, isolation, and eventual geographic adaptations, (Rosenberger et al, 1991; Rosenberger 1992;Ciochon et al, 1980; Fleagle, 1999;Houle, 1999). The geographic questions appear to be the main ones surrounding platyrrhine origins. Such issues require the examination of both ecological information as well as what little fossil information exists.

South America was an island continent throughout most of the Cenozoic, separated from Africa and North America. Most studies indicate that positions of North and South America and Africa were much as they are now during the Eocene and Oligocene in terms of position, this is due to the fact that rifting had taken place much earlier during the Mesozoic. During the Cenozoic there were thought to be some large areas of shallow waters resulting from crustal uplifts and the possible existence of several islands throughout the southern Atlantic. In other words, much dry land was available along the continents. Most evidence tends to favor a crossing of the waters via "rafts" of vegetative or land matter from Africa to South America. Wind and water currents at the time favored "floating" from Africa. Most likely, in order to have survived such a journey, these primates had to have been pre-adapted to strong seasonal variations in water availability. This is quite possible, as in Africa many climate changes (warm to cold and back to warmer) affected vegetation and water, particularly during transformations from more forested regions to the expansion of the savannahs, (Simpson 1980;Rosenberger et al,

1991; Rosenberger 1992; Fleagle, 1999). This would have been taking place during the times of the early evolutionary stages of primate evolution making it necessary to be able to adapt to ever changing conditions over time. So, it is likely that the migratory primates were able to recognize dry seasons and were able to utilize alternate food sources until reaching land. This in turn could have provided them with the ability to adapt to a variety of dietary niches upon their arrival as well, leading to today's diversification and modern forms. Africa, therefore, has turned out to be the most likely source of early platyrrhines, as the only true anthropoids have balso been placed in Africa during the Oligocene, and there have been found to be a number of similarities between those fossils found in Fayem Valley Egypt and both fossil and extant platyrrines. Parapathecids have been found to be the most likely predecessors to the divergence of both catarrhines and platyrrhines based upon cranium and dentition similarities. The similarities between South American rodents has also been noted as providing additional faunal support for the "rafting" hypothesis, (Simpson 1980; Rosenberger 1992; Fleagle, 1999). One or more species of *Branisella* most likely arrived and adaptive radiation occurred leading to diversification and continual adaptations to given environments and dietary niches throughout the continent.

Branisella dates to about 26 million years ago and is known (as is much of the fossil material) from only dental and jaw fragments, yet it is the most likely candidate to serve as the link between African origins and the appearance of platyrrhines in South America. As a result of continental drift, prosimians or additional anthropoids then were mot likely prevented from reaching either South America or Australia, leaving the platyrrhines to develop an independent and distinct pattern. The adaptive radiation of the platyrrhines began to unfold as major lineages became diversified within specific ecological/dietary niches, (Simpson 1980; Rosenberger 1992; Fleagle, 1999). The available fossil evidence tends to lend support to the idea of the African origin and the adaptation to specific niches.

The earliest connections platyrrhines have to Africa lie in the fossils of *Branisella*, found in Bolivia along with *Szalatavus*. These both have been determined to be dentally similar to modern day tamarins and marmosets, while also demonstrating a relationship to the African fossils from the El Fayum Valley in Egypt. During the early to mid 1990's more *Branisella* fossils surfaced in Bolivia, again dental. The new premolars found were found to be related to Callitrichines along with the *Szalatavus* specimens. In 1995 a complete platyrrhine skull was found in the Andeas of Central Chile, which due to its pereservation serves as the best indicator thus far of an African origin. A primate scapula and ulna fragments recovered in Argentina in 1991 were found to not only resemble living platyrrhines such as Cebus, but given the arm length and breadth, it was also determined this species may have led to branching of what are now the arboreal *Alouatta* and *Lagothrix* genera. One indication of expansion and the adaptive diversity of the platyrrhines showing they clearly underwent body size expansion during the Pleistocene was the find of a nearly complete skeleton of a robust body type resembling living spider monkeys. It was found in Pleistocene deposits in Brazil in 1993. The skeleton showed the highly specialized post cranial pattern shared by spider and wolly monkeys exclusively. The specimen suggests that they were once twice the size they are today. In 1986 anatomical evidence from additional skulls and dentition from Colombia showed yet another clear transitional link between Neogene fossils and the owl monkey of today. All fossil evidence has tended to show extensive similarities with modern genera, (Simons, 1976;Ciochon et al, 1980; Fleagle, 1999;Houle, 1999).

All of the fossil species of platyrrines have thus far been shown to be potentially good candidates as ancestors of their extant relatives. The La Ventan *Aotus* is further support for the idea that there indeed was a modern platyrrhine radiation which includes many species traceable to the early Miocene, while also providing data in support of an African origin. Both fossil data, though rare, along with data pertaining to continental drift and the available wind and water current data at the time are highly supportive of not only an African origin of today's platyrrhines, but also of their divergences from earlier forms showing environmental/dietary niche adaptation along the way to modern forms.

Hominoids

Miocene Radiation of the Apes

During the Miocene, about 23-5.5 million years ago, (Rosen, 1974), there was an extensive diversification of ape-lie creatures. Miocene apes exhibited a wide range of diversity with respect to shape, size and distribution. They ahve been found throughout Europe, Africa, and Asia. The earliest ones , however, originate from Africa. During the Oligocene, Africa was a lowland tropical rainforest, but that all changed come the Miocene. The early Miocene opened the way for a more diversified environment to flourish. This challeneged the stem catarrhines , and forced them to adapt to new habitats. Partly, greater fluctuations in climates during the seasons led to greater variation in conditions even in any given single locale. More importantly, though, areas of grassland and savanna expanded at the expense of the rainforests. The climate was changing, in part, due to the fact the continent of Africa was drifting and moving northward. At the begining of the Miocene, Africa and Arabia formed a southern island continent which was isolated from Eurasia by the Tethys Sea. Thus, the Atlantic and the Indian Oceas were connected. The Mediterranean is now currently what is left of what was once the Tethys. The geological action, continental drift, brought about by plate tectonics as previously discussed , caused a collision between the southern and northern continents approximately 17 million years ago, which resulted in new geological features and therefore new climatic patterns. For the apes, it had important effects in their overall adaptations and colonization patterns, and in areas which previously did not have any such animals residing in it. The result then of this environmental diversity was the emergence and diversification biologically among what are known as the Miocene hominoids,(McHenry, 1975;Howell, 1977;Pilbeam, 1986; Skelton et al, 1986; Simons, 1989).

Early Miocene hominoids are found only in Africa, later Miocene hominoids are found throughout the Old World. Of the latest Miocene hominoid species, which appear to have been fewer in number than their predecessors, one group became adapted for bipedal locomotion and terrestrial life habitually. Some have attempted in the past to group the Miocene hominoids into two groups, those ancestral to humans and those ancestral to the other extant apes. However, it is now known that few fall into such catagories, and there are also so few moderns among them. Those which became extinct also did not leave living descendents for the most part and it is difficult to decipher which are the few who did. What will follow will proceed to discuss some of the primary Miocene hominoids and their relationships to the overall process of human (primate) evolution , while proceeding to uncover the difficulties in understanding the roots of the Family Hominidae, (1975;Howell, 1977;Pilbeam, 1986; Skelton et al, 1986).

Miocene Paleoecology

The Miocene epoch began approximately 23.7 million years ago and lasted approximately 18.4 million years (Harland et al, 1990; Park, 1999). It was originally dated to approximately 12 million years ago until the development and use of argon/argon dating (K-40/K-39 ratio) proved to be reliable enough to determine the detailed thermal history of rock (Harland et al,

1990; Dodd et al, 1990). This is considered an evolutionary phase in which mammals began to increase in number and spread throughout the Old World (Staski et al, 1992; Park 1999). There have been significant geological events (biotic events) which reflect the changes in both the paleogeography and climate that are the responsible factors for the development of the evolutionary pattern followed by both plants and animals, and their ecosystems (Foster et al, 1970; Aubry et al, 1988).

The Miocene is considered to be one of the largest transitional times, a time when more modern-type ecosystems began to emerge. The overall pattern of biological change is one of expanding open vegetation systems (deserts, tundras, and grasslands) and a decrease in forests. This led to a rediversification of more temperate ecosystems and numerous morphological changes in animals, (developing herbivores, predators, rodents), and an increase in more modern vegetation (as evident by the many pollen and spore studies conducted). A mid-Miocene warming , followed by a cooling, is considered responsible for the retreat of tropical ecosystems, the expansion of northern coniferous forest and increased seasonality. Kelp forests also appeared for the first time (Hayes et al, 1976; Aubrey , 1988; Pole, 1993).

During the late Miocene the island continent of India slammed into Asia, thereby pushing up the Himalayas and triggering a global cooling that would last into the Pleistocene, the Rockies and Andes also rose in similar fashions. Throughout the early and mid- Miocene however, the climate was generally warmer globally. It was during this time that increasingly more modern patterns of atmospheric and ocean circulation formed. Antarctica became isolated from Australia and South America establishing the polar ocean circulation, eventually leading to the ice cap build up. The most significant events of the Miocene were the formation of the East Antarctic ice sheet, closure of the equatorial Indo-Pacific passage, and the development of specific air and ocean circulation patterns throughout the Mediterranean. It was such cooling and drying during the mid-Miocene and late Miocene which was responsible for the expansion of the grasslands, the rise of mammals, especially anthropoid apes, and the start of massive migrations on a global scale (Hayes et al, 1976; Aubrey, 1988; Pole, 1993; Park, 1999).

Perhaps in terms of the evolutionary and taxonomic status of both plants and animals however was the fact that 40AR/39Ar dating of strata extended the Miocene by nearly 7 million years. This is a radioisotopic dating method that entails measuring the ratio of two argon isotopes, argon-39 and argon-40. Measurements of both isotopes are simultaneously made at the same location in the crystal lattice where the argon is trapped (Alvarez et al, 1997; Staski et al, 1992). It has been proven to be a reliable dating method, which in turn has made the study of the Miocene far more interesting in terms of the many avenues open to reexamination and reinterpretation overall.

Lothagam Mandible

Though most of the literature places the Lothagam site in Northern Kenya, just west of Lake Turkana (Rudolfensis), it would appear that Lothagam itself actually is found to be located in Uganda, just over the Kenyan border. So, most likely this is due to the possibility that the site has extended over the border. The Lothagam area itself has been described as being a dry, desert-like plain region in the north (at an altitude of 1,329 meters) comprised of a slight westerly slopped sequence of volcanic and sedimentary rocks. Its coordinates, 3, 1'0N by 34, 28' 0E, place it just over the Northern Kenyan border in Uganda. The lower part of the rock sequence, lavas, and the more course volcanic sediments of what is know as the Nabwal Arangan beds, is said to have been deposited between 14-12 million years ago, while the upper sections are dated to about 9 million years ago. What are believed to be two hominoid teeth found in the higher formation has been dated to between 6.5-5 million years ago, and the Lothagam mandible still in question today is said to be more than 4.2 million, but younger than 5 million years ago, (Hill et al, 1992; McDougal et al, 1999; calle.com, map 2000).

Since the Patterson 1967 team discovery of the Lothagam mandible, hundreds of bone fragments and teeth were uncovered between 1990 and 1993from primarily the upper portion of the lower section of the sediment strata (Miocene-Pliocene region). Most have been dated to be slightly younger than 4 million years ago. The Lothagam mandible has been presumed to be the earliest hominid fragment dated to about 5.7 million years ago. This site has been a major contributor to the revisions continually being made in both stratigraphic frameworks and interpretations of paleoenvironmental settings. The sediment layers suggest a mid to late major Miocene environmental change, (Hill et al, 1992; Leakey et al, 1996; Stewart, 1997; McDougal et al, 1999).

The 5.7 million year mandible is also part of the reason, as it has sparked debate over the origins of hominid species, as well as over it's own identification and classification since it has been argued to belong to either *Ardipithecus ramidus* (dating to about 4.4mya) or *Australopithecus afarensis* (dating to about 4.2mya). Since the Lothagam fossil dates further back, and appears to have some hominid characteristics, it has been tossed between these two. However, it has yet to be determined if it is hominine or panine, making it a fossil which would actually resemble hominoid dentition instead.

Ardipithecus ramidus though is the oldest recognized hominid species thus far, (dated to just after the hominine/panine phyletic split), and since the status of the Lothagam fossil is still questionable yet dates to before this and appears quite similar morphologically, many have chosen to place it with *A. ramidus*. This was due to the fact that this particular species is thought to be an ideal intermediate between *A. anamensis* and *afarensis*. Dentally, compared to later hominids, it has a larger canine in comparison to the post-canine, smaller first premolars, and a thinner enamel. When compared to African apes the canines are smaller, and the enamel is thicker, with more molar-like premolars. It is also non-sectorial as opposed to sectorial as found in the pongids. Therefore, since the tiny mandibular segment found at Lotahgam appears more hominid-like and less pongid-like, it has been tossed between *ramidus* and *anamensis*, while seemingly better suited to *ramidus,* (White, 1980; Hill et al, 1992; Leakey et al, 1996; Stewart, 1997; McDougal et al, 1999).

The paleoenvironment of the time has been fairly well preserved and has been interpreted as being indicative of a forested setting, as supported by macrofaunal remains. Some Colobines and baboons were known to be present, (White, 1980; Hill et al, 1992; Leakey et al, 1996). The approximately 5.7 million year old mandible from Lothagam, similar morphologically to *Ardipithecus ramidus* at 4.4 million years ago, in conjunction with the environmental data suggests that neither may have differed all that much from modern African apes, and that Lothagam may very well have been an early point on the hominid lineage.

Proconsul

Proconsul is one of the best represented Miocene hominoids in the fossil record, dating anywhere from approximately 23 to 14 million years ago. From numerous examinations of the remains, it appears that there is considerable variation within the genus. Now one of the best known of the ape ancestors, and considered the last common ancestor of both humans and apes, the fossilized remains were first unearthed in 1927 in Western Kenya by a settler named H.L. Gordon, and first examined and described by Tindell Hopwood (1933). Though the relationship between many of the Miocene apes and anatomically modern hominoids is still not well understood, a good many of the Miocene specimens have been placed within the family *Proconsulidae*. *Proconsul* appears to be close morphologically, particularly on the basis of dentition , to lines which lead to all later hominoids. It also may be that later, established, genera such as *Drypithecus* and *Sivapithecus* more appropriately belong re-classified into *Proconsulidae* as opposed to *Pongidae* as species within *Proconsul*. *Proconsul* apes, *africanus* in particular , have been perceived of as being "dentally specialized" (highly adapative dentition for the varied environments of the early to mid Miocene). It is this particular feature which raises questions about the role of *Proconsul* in hominoid evolution and it's relationship to other Miocene hominoids (Simmons, 1972; Schwartz et al, 1978; Szalay et al , 1979; Fleagle, 1988).

Early on, during the rise of the hominoids, there is a split which begins between cercopithecoids and hominoids, brining about the appearance of *Proconsul* about 23 million years ago. This split is characterized by the following features: Y-5 molar pattern (the cusp pattern found among humans and is a characteristic found among early hominoids), loss of tail, larger brain size, and a more mobile shoulder and elbow joint for suspension. Granted *Proconsul* has a more monkey – like posture as opposed to the trunks of more anatomically modern apes, but this is due to the belief that since great variation existed, many inhabited a wide range of environmental niches and adapted accordingly. Body sizes also ranged from about 10 to 150 pounds. The dental formula was similar to that of the Old World anthropoids, 2-1-2-3, and their enamel thin suggesting their diet consisted of fruits and vegetation primarily. Given the similarities (primarily dental) among *Proconsul* and others such as *Dryopithecus, Sivapithecus,* and the dentition of anatomically modern humans, it is likely that this genus was not only the result of the split between hominoids and cercopithecoids, but was also responsible for the *Dryopithecine* lineage which led later to other apes such as the Gibbons and Siamangs (Hylobatidae), and larger bodied *Sivapithecines* which led later to Orangutans (Pongidae) and the smaller *Sivapithecines* then leading to *Autralopithecines*, (Hominidae), (Simmons, 1972; Scott et al, 1974; Schwartz et al, 1978; Szalay et al , 1979; Fleagle, 1988).

Since most of the Miocene Hominoids belong to Proconsul, it could perhaps be argued based upon morphological comparisons (primarily dental) that the Miocene hominoids appear actually to belong to one genus, *Proconsul*, with many species given the variation within the genus and the variation in environmental niches. Given the dental similarities alone, it could also be argued that both *Sivapithecus* and *Dryopithecus* would best be suited to placements as species within this genus. The common dental formula has changed very little, as well as cusp pattern and enamel. The most likely conclusion would be that this one genus displayed such variety that it not only was comprised of numerous species, but also did give rise to three other main family lineages, (Simmons, 1972; Scott et al, 1974; Schwartz et al, 1978; Szalay et al , 1979; Fleagle, 1988).

Sivapithecus

In terms of an approach to take when studying such Miocene Hominoids as *Sivapithecus*, there are actually two ways in which to view it. One would be from a dental perspective, the other from the paleoenvironment. In looking back at where such a species perhaps originated, it appears as though there are perhaps two possibilities. In looking at the remains of *Proconsul (Proconsulidae)*, it is believed that this is the origin point for other hominoids of the mid to late Miocene. When comparing the remains of this genus to those of *Dryopithecus* and *Sivapithecus*, craniofacial traits which are found in all, including later apes and Australopithecines can be seen. In terms of taxonomic status, it is possible that *Sivapithecus* then was either a derivative of *Proconsul,* or perhaps even a variation of one of the *Dryopithecines*. It is well documented that a great deal of variation existed within both the genus *Proconsul,* and one of it's species (most likely, based upon fossil evidence), *Dryopithecus*. All of these hominoids also are known for having varied environmental habitats, wide-ranges, and therefore the probabilities of seeing the development of local adaptations would be increased. This would result in the high degree of intra-species variation being reported in the literature. It would also perhaps offer an explanation for the divergence patterns of Miocene hominoids overall. In terms of *Sivapithecus,* it too displays the Y-5 molar cusp pattern found in anatomically modern humans. It is possible to think that both *Sivapithecus* and *Dryopithecus* are best suited to placement as species within the genus *Proconsul.* From dental and cranial images, it appears that *Sivapithecus* gave rise to not only present day Orangutans, but also to the Australopithecine lineages, while *Dryopithecus* possibly is responsible for present day gorillas and chimps, (Simons et al, 1965; Martin, 1985; Pilbeam et al 1990; Beyon et al, 1991; Grine, 1991; Benefit et al, 1995; Scott et al, 1999).

Martin (1985) reviewed the evolutionary significance in hominoid evolution by studying cut faces of hominoid teeth using various methods of analysis in order to relate enamel prism patterns and thickness. He looked at formation rates, and found that thick (fast rate) enamel was found to be ancestral to humans and great apes, while thin (slow rate) enamel was common to African apes. This data could be interpreted as support for the variability in *Sivapithecus,* (as well as other Miocene hominoids), and could also explain how the wide range of environmental habitats led to localized adaptations, again to the variability, and the possibility that this variability led to a divergence of other lineages off of one species. In the case of *Sivapithecus* this would mean it could have been possible that it led to both Orangutans and Australopithecines. Studies have also linked various morphological traits such as size, femoral characteristics, and craniofacial features to the environments. The thickness of the enamel and rates of formation, size of the individual, and limb type adaptations would therefore also be localized adaptations in terms not only of the environment, but dietary factors as well (which are also somewhat rooted in the specific environments). So, depending upon the degree of vegetation present in an environment, evolutionary patterns would be affected in both teeth and bone in general. This would account for the high degree of intra-species variation once again.

Sivapithecus also displays the 2-1-2-3 dental formula, along with sexual dimorphism in size of teeth (some literature still places *Ramapithecus* within the same genus as a female version, despite some morphological differences in both the teeth and cranium). *Sivapithecus* displays more Orang-like features, though depending upon the specimen being examined, could also lead to thinking it also displays early hominine features. The conclusions drawn would be that this species most likely evolved from *Proconsul,* more specifically , from the *Dryopithecines* it would seem, and due to the intra-species variations in both morphology and environments it most likely led to a divergence within the species bringing about both Orangutans and Australopithecines, (Simons et al, 1965; Martin, 1985; Pilbeam et al 1990; Beyon et al, 1991; Grine, 1991; Benefit et al, 1995; Scott et al, 1999).

Dryopithecus

The *Dryopithecines* are a small group of generalized apes which can be thought of as a group which led to both anatomically modern apes and humans. Although *Dryopithecus* has been known by a variety of names based upon fragmentary material found over a widespread area (Europe, Asia, Africa), it appears probable that despite it's classification in by Lartet in 1856 as a genus within the family *Pongidae,* that it perhaps is better suited to the status of species within the family *Proconsulidae.* It is quite possible that *Proconsul* not only gave rise to *Pongidae* lineages, but even more possible that due to many dental similarities and given that the literature regarding both *Proconsul* and *Dryopithecus* clearly indicate that there was considerable variability based upon the fossil evidence of each, that *Dryopithecus* indeed could be a species of *Proconsul.* The *Dryopithecines* are remarkably similar in terms of skull and dental comparisons to *Pongids*, (gorillas in particular), leading to the opinion that they did in fact develop within *Proconsulidae* and during the course of future divergences and adaptive processes, (as they too occupied a variety of ecological niches), led to the emergence of the *Pongids. Sivapithecines* also perhaps deserve similar taxonomic reconsideration, but yet seem to resemble more the Orangutans within *Pongidae*, yet also displays some morphological similarities to early *Australopithecines* (particularly from the perspective of dentition). This leads to the opinion that despite it's original classification in 1910 by Pilgrim, that it too is perhaps best placed as a species under *Proconsul*, and was also part of a divergence that perhaps occurred with the *Dryopithecines* (it leading to Orangutans and Austalopithecines, and the *Dryopiths* leading to other apes). All three tend to be depicted in the literature as having displayed a great deal of variation within their respective groups, as well as having inhabited a wide variety of niches. *Dryopithecus* therefore most likely gave rise to a derivative that eventually led to the earliest of hominine forms, and given the remarkable resemblance even to a close branch such as *Ramapithecus* (thought to be an early hominid by some), the possibilities may also exist that they also gave rise to *Sivapith* lineages as well. So, despite the literature and some issues surrounding dates of naming of these specimens, it is still felt that there is enough evidence to support consideration of *Dryopithecus* as a species within the genus *Proconsul* and within the family *Proconsulidae,* (Holloway, 1967; Simmons, 1972; Pilbeam, 1978, 1986; Szalay et al, 1979; Beynon et al, 1988; Grine, 1987; Anemone et al, 1990; Begun, 1994; Andrews et al, 1996).

Both environmental and cultural (localized) adaptations are other factors which must be considered when examining the placement of not only *Dryopithecus*, but all Miocene hominoids in general. The dental formula of *Dryopitjhecus* is still in line with other hominoids, 2-1-2-3, with sexual dimorphism expressed in tooth size (canines in particular) and the P3 sectorial presence. The shape of the maxilla or mandibular pieces is suggestive of perhaps an intermediate form, as it is not really a parabola as found in hominids, but yet not quite as squared off as found in *Pongids*. This point may be key to understanding the relationship *Dryoptihecus* actually holds with other Miocene hominoids overall then. Zygomatic characteristics and phylogenetic evidence also would tend to suggest that there is more of a relationship with gorillas than with orangutans.

Therefore, the conclusions drawn must lead to pointing towards *Dryopithecus* as a species within *Proconsul* (*Proconsulidae)*, and a predecessor even perhaps to *Sivapithecus* lineages, other *Pongids*, and eventually early hominines. Given again the high degree of variability and the numerous ecological niches inhabited, there is perhaps some reason to believe this to be possible, (Holloway, 1967; Simmons, 1972; Pilbeam, 1978, 1986; Szalay et al, 1979; Beynon et al, 1988; Grine, 1987; Anemone et al, 1990; Begun, 1994; Andrews et al, 1996).

Ramapithecus

During the 1960's Elwyn Simons and David Pilbeam, (then at Yale University) began a campaign to bring life back to a primate of the Miocene, *Ramapithecus*. Named in 1932 by G. E. Lewis, *Ramapithecus brevirostris* was an ape-like primate having a short face and fairly small canines. It was a species primarily only known from tooth fragments, and yet it appeared at first to be more like humans and less like chimpanzees in a number of significant ways, (Andrews et al, 1982; Lewin 1988; Staski et al 1992).

Not only were the canines small and the face short, *Ramapithecus* was found to have thick enamel on its molar teeth, again more human-like than chimp-like. Furtehrmore, it also seemed to have a tooth-row which diverged at the rear, (the more parabola shape common to apes and humans), not parallel (u-shaped as in chimps). Simons and Pilbeam felt that Miocene primates were taxonomically "oversplit", meaning too many names/species for what could be actually justified by the fossil record, and therefore combined the fossils of both *Kenyapithecus* in with those of *Ramapithecus,* justifying a new human ancestor, (Andrews et al, 1982; Lewin 1988).

Due to this now more specific alliance with humans, and also due to the inclusion of the fossils of *Kenyapithecus,* the age of *Ramapithecus* was forced to be dated then to about 14mya, the known age of *Kenyapithecus.* It was on this basis that Simon and Pilbeam argued for the split of humans and apes to be pushed back to prior to 14mya, when they claimed now that the first primate unlike a chimp, and more like a human, based on their dentition analysis, could be detected, (Andrews et al, 1982; Lewin 1988). There were several problems with their thinking however.

First, they assumed that similarity was an indicator of the proximity of descent. This, however, tends to lead to a degree of morphological confusion regarding perceived derived characteristics possessed by several species which would indicate inheritance from a recent common ancestor. Second, the then assumed that the diversity of the apes in the Miocene was comparable to the diversity of apes alive during recent times, not taking into account changes in DNA over time (phylogenetic changes). Third, and perhaps the most significant of the problems, was the fact that they implicitly divided the fossils into two catagories which they referred to as " us/not us" or simply "human/ape". This tended to imply that the apes were all more closely related to one another than to humans, and that all the evolutionary changes occurred in the human lineage alone. Therefore, it was stated that humans had to be full of derived features, and any similarities between a human and a Miocene ape were considered to be a synapomorphy (derived characters possessed by several species indicating inheritance from a common ancestor). Their claim was that these characteristics could be detected in the fossil record as early as 14 million years ago, (Lewin 1984, 1988; Thorne et al 1992).

This interpretation was challenged by studies of molecular evolutionary rated during the late 1960's. If molecular data were being interpreted properly, then the 14 million year date for the divergence of humans and apes might very well have come out to be a serious overestimation. As it would turn out, humans, chimps and gorillas were so genetically similar that it was considered nearly impossible for them to have been separate lineages for that length of time. Subsequent discoveries in Turkey Pakistan showed that *Ramapithecus* was not all that similar after all to either humans or chimps, but rather to the orangutan. The thicker molar enamel was also found to exist in orangutans. The evolutionary event in question, it would seem, was the issue of the thinning of enamel, since it occurred in all the common ancestry of chimps and gorillas. Other characteristics, such as the parabolic dental form were also reconsidered. These new finds suggested that Simons and Pilbeam had misinterpreted the old fossils (Andrews et al 1982; Lewin 1984, 1988; Thorne et al 1992).

Furthermore, the Miocene apes have come to be recognized as having been exceedingly diverse, moreso even than contemporary apes. Only one species at any time has been thought to be part of human ancestry, and only a few species are even in the ancestry of all living apes; most have become extinct without ever having left any evidence even of their existence. This means basically that it is nearly impossible to actually isolate the specific one which gave rise to humans, and it may never be found. Grouping all apes that were not human-like into a single category created a phylogenetic heterogeneous group. By assuming that anything like humans is therefore a close relation and therefore anything not like humans is closely related to anything not at all like humans they failed to differentiate between similarity and synapomorphy. Simons and Pilbeam were making much weaker phylogenetic inferences than they thought. *Ramapithecus* is now widely accepted to the be female *Sivapithecus*, and both are closely related to the orangs. *Kenyapithecus* is thought to be a genus all its own again, and it may or may not ever emerge as ancestral to some or all of the living apes and humans, (Andrews et al 1982; Lewin 1984, 1988; Thorne et al 1992).

The *Ramapithecus* discovery demostrated three important issues. First., it highlighted the difficulty in inferring ancestral/descendent relationships from the fossil record. Second, it illustrates the incompleteness and interpretive difficulties of the fossil evidence. Finally, it shows how genetic data can be used to test hypotheses of the evolutionary relationships of organisms, yet how such evidence is intricately tied to morphological interpretation; one can not be utilized to justify relatedness independent of the other.

Aegyptopithecus

Perhaps the best candidate for an early ancestor of the great apes is a primate from the Fayum Depression of Egypt. Found by Simons (1972), *Agyptopithecus zeuxis* is known from one skull (without the mandible) and some jaw fragments. It dates from the late Oligocene (28-30mya), predating *Dryopithecus* of the Miocene. What is known of its total morphological pattern turns out to be significant among the hominoids, (Simons 1970, 1972;Olson, 1981).

Aegyptopithecus had the Y-5 molar cusp pattern, and is the oldest known fossil primate to posses it. The Y-5 cusp pattern is characteristic, as previously noted in the dryopiths, of hominoids which later gave rise to the hominid lineages. It also had large, projecting conical canine teeth, yet still retained a prosimian-like muzzle. The remainder of the skull is reminiscent of both the monkeys and lemurs. A small and weakly developed sagital crest was found as well. The supraorbital region was moderately developed into supraorbital ridges, while the back of the skull had a well -developed occipital crest. What is known of the post cranial anatomy indicates that *Aegyptopithecus* was a monkey sized primate who possessed a tail and was most likely a generalized quadruped, given the fact that the Fayum at the time was highly tropical. It makes for an excellent structural ancestor to *Dryopithecus,* and may be better placed taxonomically within the dryopith genus as *Dryopithecus zeuxis,* (Day, 1970;Simons 1970, 1972;Olson, 1981).

Gigantopithecus

Over the course of many centuries oriental pharmacologists have marketed ground up fossil bone and teeth as something known as "dragon bones", and promoted these as cures for a variety of ailments. In 1935, G.H. Ralph von Keonigswald found one such tooth, a molar belonging to some type of fossil hominoid. He believed it to be a pongid , and named it *Gigantopithecus.* By 1939 he had found other similar molars , but no jaws or other skeletal remains to coincide with his other teeth. These teeth are believed to be of the middle Pleistocene and were found in fossil beds in Southern China, making *Gigantopithecus* a leader as a contemporary of the true fossil hominids to come, (LeGros Clark 1954, 1959, 1965;Howell, 1967; Day, 1970; Simons,1972).

Over the course of the 1950's the first discoveries of what were *Gigantopithecus's* mandibles were recovered in China, known as the Kwangsi mandibles. In 1968 another mandible was found in India which dated to around the middle Pleistocene, indicative of populations which probably existed from about 5 million years ago to possibly 250,000 years ago (Birdsell, 1972). The early part of the Pleistocene is known for megafauna (large animals which are today much smaller). The gigantopithecines of the Pleistocene likely represent one such group with respect to extant gorillas. There were also some known "giant" baboons in east Africa at around 2 million years ago (Birdsell, 1972) existing at about the same time as the australopithecines. The Pliocene *Gigantopithecus* jaw is not all that large, and may be a female, indicating the presence of sexual dimorphism within the populations, and is confirmed in the Chinese specimens, (1965;Howell, 1967; Day, 1970; Simons,1972).

The largest *Gigantopithecus* mandible actually is smaller than that of extant gorillas, so caution is best when assessing total body size on the basis of teeth and jaws alone. For it is also known that the australopithecines possessed larger teeth and jaws than anatomically modern humans, yet were quite smaller in terms of total body size. Mosaic evolution , therefore, seems to have always been ongoing with respect to the higher primates , (Simpson, 1949). It has been suggested by Simons, (1970), that *Gigantopithecus* weighed nearly 600 pounds and was perhaps over nine feet tall, though this has yet to be verifiable. It is highly likely though that these large simian shelved mandibles did indeed belong to gorilla-like primates who' s evolutionary lineage has yet to be determined, though they may well be "dead ends", in addition to being descendants of the dryopithecines. *Gigantopithecus* canines were not excessively projecting or large. Its molar teeth were in some ways quite hominid. It was such features which led Franz weidenreich to deduce that this primate was in fact some type of giant hominid, which he named "Giganthropus". He believed that man's early ancestors were giants. This idea had been well refuted by fossil evidence though. At present, *Gigantopithecus* is considered a giant gorilla-like pongid who was terrestrial and herbivorous, which possessed some early hominid traits. It could very well have been one lineage which arose from the dryopiths , as it is evident in the fossil record that they were a highly generalized group of hominoids which followed a number of evolutionary pathways, (LeGros Clark 1954, 1959, 1965;Howell, 1967; Day, 1970; Simons,1972).

Oreopithecus

Many have considered it another evolutionary dead end, while others place it within the Cercopithecidae, and still others who believe this was one of the earliest hominid forms capable of habitual bipedal locomotion, (Simons, 1972; Delson, 1981; Harrison, 1986; Fleagle, 1999). It is quite possible though that *Oreopithecus bamboli* , first found in Tuscany, Italy in 1872, in what was once swamp and now a coal mining area, is an example of one of the first true hominids. Though also known as the "swamp ape", few consider it an ape, (Simons, 1964).

Oreopithecus is known from some fifty individual specimens found from 1954 on by Johannes Hurzeler, and dates from the late Miocene, about 10 million years ago. It presents the most unusual mosaic of primate traits to date. Its brain was about the size of a chimp's, but the skull has a non pongid -like appearance (well rounded cranial vault with large supraorbital ridges) (Simons, 1964). The cranial vault is well rounded and smooth, similar to monkeys. The heavy supraorbital torus overhangs the quite vertical and shortened face. There is also a well developed hominid nasal bridge. The canine teeth project, but are not considered large in comparison to those of apes and monkeys. The incisors are small and more vertical, like hominids. The entire dentition in fact is balanced (hominid trait) in which the anterior and posterior teeth are proportionate in size to one another. Diastamata do not regularly occur, nor does the lower jaw possess a simian shelf. The upper molar teeth show some similarities with the cercopiths, indicating some ancestral affinities perhaps, and lineage divergences, but a more hominid like creature than any cercopith. However, more evidence to support such a divergence can be found in the 1920's work of William Gregory that noted the cheek teeth of the lower jaw and their strong resemblance of *Apidium's,* but no direct ancestry is supported or sufficient enough to place *Oreopithcus* with the cercoptihs due to the fact that a direct relationship is ruled out on the basis of *Apidium's* lacking of one pair of incisors still present in *Oreopithecus,* (Simons,1970). The skull base is more crushed and may be indicative of a more forwardly placed foramen magnum, as well as more forwardly placed occipital condyles. The upper limb of *Oreopithecus* is quite long, similar to a brachiator, yet the iliac blade portion of the pelvis is horizontally expanded in a hominid manner. This tends to indicate an erect posture and bipedal locomotion, yet still maintaining a partial brachiating habit while in the forest swamp regions.

Some foot bones also exhibit hominid traits, yet it still had an opposable large toe, (Simons, 1970; Pilbeam, 1972; Simons, 1972; Harrison, 1986: Delson, 2000).

Few still classify *Oreopithecus* as cercopithecoid, pongid or hominid though. The majority tend to favor placing this primate in the superfamily *Hominoidea*. While Hurzeler and William Straus, the two most noted for study of this particular primate, tend towards placing it as an aberrant hominid, while others still tend to want to give it its' own family, Oreopithecidae. If *Oreopithecus* is so unusual, what then makes it that important? It is because this primate establishes and makes evident that hominid, pongid and cercopithecoid traits were probably acquired (as well as lost) independently a number of times in the course of primate evolution. *Oreopithecus* further clarifies the principle that no one animal can be defined or classified by isolated single traits , only by total morphological patterns, (LeGros Clark, 1959). It is evident that during the Miocene and into the early part of the Pliocen, higher primate populations were evolving along a number of adaptive pathways with some remaining generalized arboreal quadrupeds and others becoming true brachiators. Finally, some were also evolving along the hominid pathway, while a few like *Oreopithecus,* were adapting into multi niches, (Huthcinson,1965). Such adaptation would likely have required a more complex neurological organization of the brain than ever before in any other fossil primates of the time. It is such potential ancestors as this then that would continually make each group of fossil primates appear more aberrant when compared to more recent evolving populations and /or extants despite the fact they could not be defined clearly as ape or monkey in terms of ancestry in any clear sense due to niche adaptations over time , thereby explaining the more generalized anatomies found in the fossil record of what are considered intermediate species, driving home again the significance of LeGros Clark's emphasis (1959) of the importance of total morphological patterns and Huchinson' s (1965) emphasis on niche adaptation over time as a continual and ever changing dynamic process in response to fluctuating environments, (Simons, 1970; Pilbeam, 1972; Simons, 1972; Harrison, 1986: Delson, 2000).

Afropithecus

There are several wide ranging perspectives pertaining to the ancestral ties of *Afropithecus*. The longer snout , straighter craniofacial profile, and smaller frontal bone have suggested to some (Simons, 1987) that this is a genus more closely related to early cattarrhines such as *Aegyptopithecus.* Yet it's specimens posses premolars which are indicative of some more advanced forms of other proconsulids , making it a genus closer in relation to the origin lines of apes and humans, (Andres *et al*, 1987, 1992).

Morotopithecus

This newest species, *Morotopithecus bishopi*, (Gebo *et al*, 1997), dates to the early Miocene, roughly 20.6 million years , though some evidence has been put forth stating that the beds surrounding the fossils were contaminated to some extent and were therefore not well enough preserved, making the date roughly 15-17 million years, as Andrews (1992, 1996) had previously dated the fauna in the Moroto regions to 17 million years, (Delson, 2000), and Martin Pickford (1999) has also suggested an age of the Middle Miocene of 15-17 million years.

In 1961, W.W. Bishop and David Allbrook took a team into Uganda and collected the craniofacial fragments which were later used to reconstruct the skull of what was to become taxonomically shifted to *Dryopithecus major,*moved by Simons and Pilbeam (1965) from *Proconsul major* on the basis of primarily dental remains, ie: the Moroto palate and the loose teeth fit into it over time. It was this palate , located in a site known as Moroto II in Uganda that is seemingly taxonomically mysterious with respect to *bishopi.*

Going back to 1969, David Pilbeam's dissertation, he was hesitant it appears, to split the dryopiths much at all, or not at all, stating that the material is highly unlikely to be that of more than one species. This is most likely the case, leaving the taxonomic creation of *bishopi* somewhat questionable with respect to the moving of what were once considered *D. major* pieces forward in time grouped now with new finds of a species most likely the same, as also noted by Pilbeam in 1969 with respect to the then issues surrounding *Proconsul major* and the move he and Simons made in 1965 of the palate and craniofacial fragments to *D. major.* The latest Moroto pieces from 1997 (Gebo *et al*) and 2000 (McClatchy *et al*) do indicate that this species, being given the new name of *bishopi,* morpholigcally is the same as *D. major*, which was presumably moved forward to become part of *bishopi*, creating a taxonomic puzzle.

In 1959, there was also a Moroto I site where Wilson found some of the first fossils of a larger ape, much more robust than any others known, and seemingly more hominid dentally, while also being somewhat upright in posture. The Bishop site is known as Moroto II, and in 1962, Bishop states that some of the fossils his team had uncovered did appear to represent *Proconsul.* It is important to note here that Wilson did not classify the fossils he came across, yet noted they were quite large and different from anything else known. There was the left maxilla found by the first 1961 Bishop team, and the right found by the next Allbrook team in 1962. According to the 1963 Bishop/Allbrook paper, the various loose teeth found (by Leakey 1958 and 1961) and placed into the then brought together palate, along with the mandibular and facial fragments, were also examined by LeGros Clark and Louise Leakey, and it was noted the two also concurred with Allbrook and Bishop that these were indeed *Proconsul.* However, confusion still existed, and this placement was noted in publication as being a provisional one. What was done by Gebo, Pilbeam and McClatchy was a bringing together, as they called it, of both the

Moroto I and Moroto II pieces into one new species based upon their new finds of various robust, upright, postcranial remains, (1997,2000). However, according to taxonomic standards, sinking can only be done in the reverse, to the earlier , and first named genus and species, though it is felt that one piece of a given collection can be moved forward if it indeed is that unfitting, thereby leaving the remainder of the species intact. From skull reconstructions though, it appears that even the craniofacial fragments of *D. major* were also moved forward with the palate, effectively and presumably dissolving the entire species. Given the fact that originally the pieces were placed provisionally into *Proconsul,* it was easy to understand how Simons and Pilbeam, (1965) were able to taxonomically take them and bring them forward to *Dryopithecus.* This move seemingly dissolved the provisional status, and is listed as a solid genus and species in both Pilbeam's 1969 dissertation, as well as the two book series he and Simons both put out in 1972.

In any case, the dryopiths certainly represent a transitional (intermediate) species. In which case, issues of again the idea of ring species existence, environmental factors, along with sexual dimorphisms must also be taken into account. The taxonomic side of this also should be concerned with how the pieces were seemingly shifted forward into a brand new species, when taxonomy only allows for a shift back to the first named genus and species. It is most likely that Gebo and his team (1997) did find some new pieces, but perhaps those which are dryopiths and bring new light to their inferred postition as an intermmediate to hominids. This status is based upon their molar (Y-5) cusp patters indicative of hominids, a thicker and more robust dentition overall with some similarities to *Proconsul* and early cercopiths, such as an incisive fossa which opened directly onto the floor of the nose, something also seen in hylobates. *Morotopithecus bishopi* , with all its similarities to such a species, and in conjunction with the rules of taxonomy, perhaps should have best been sunk back to *D. major*, since it's postcranial anatomy unearthed to date also indicates the same intermediate, robust, and somewhat upright hominoid along the hominid path.

Early Bipedalism

Many of the Miocene ape locales have brought a diverse return of fossil information to the surface. Fossil monkeys such as *Victoriapithecs* and fossil prosimians such as *Komba* are known from the Miocene, (Delson, 2000). The tendency to focus on these apes is most likely due to the fact that they are the most directly related and most relevant to the history of anatomically modern *Homo sapiens*. There are some hominoid fossils from the late Miocene-early Pliocene when human lineages began to emerge, and Andrew Hill along with Steven Ward (1988)concluded that only eleven African fossil specimens could that clearly be identified as being hominoid and between 14-4 million years. However, some pieces, such as teeth and mandibular fragments found in Tabarin in Kenya , are dated to 5-6 million years and are hominid in nature, though teeth are not the distinguishing feature alone between hominoids and hominids, (Delson, 2000).

During the latter part of the Miocene , a population of hominoids began habitually walking in a bipedal manner as opposed to tree suspension or quadrupedal locomotion. These were early intermmediates which became the first of the true hominids. Unfortunately the fossil record of this period i s exceedingly sparse until about 4.5-3.7 million years ago when hominoids are found already to be in possession of many of the main features of bipedalism, (Reader, 1988, Simons, 1989).

There is a definite anatomical basis for bipedal locomotion. It is also the type of locomotion pattern in the morphologies which is used to distinguish humans from apes. The first anatomically modern human ancestors were the hominids, whose adaptations for bipedalism were selected for and carried on throughout the human branch. Therefore, the hominids, (in the human family, a subcatagory of the human and ape superfamily), of roughly 3.7 million years ago possessed ape-like brains and teeth, but had the more human-like trait of walking on two legs, (Childe, 1939;Binford, 1968; Aiello *et al,* 1990).

When comparing the locomotor anatomy of apes to taht of humans, there are a number of significant differences seen, all of which are structural modifications of the skeleton. On the skull, there is the occipital condyles, which are found directly beneath, and not toward the rear, in the vertically held head of a human. Likewise, the foramen magnum (the hole which the spinal column emerges from) is directly beneath the skull in bipedal skeletons. The human vertebral column has more marked curves than that of apes, and the lumbar curve in the lower back only appears in human spines. It's purpose is to offer stronger lower back support for the overall weight of the body in an upright position. The pelvis also takes on new meaning, that of bearing the brunt of the upper body's weight, and in a bipedal specimen, the pelvis is located directly beneath the trunk. In quadrupeds it is located behind the trunk. The iliac blades have also become shortened and rotated, forming a bowl like structure. The lower limbs have also become longer and more muscular with respect to the upper limbs., and it is the lower limbs of a human which comprise more than 30 percent of the weight of a human, but less than 20 percent or so the total weight of a chimpanzee, (Olson, 1981;Aielo, *et al*, 1990). The hip joint (formed by the head of the femur inserting into the acetabulum of the pelvis) has also been strengthened and reoriented. Part of its new roles has been taken over by an expanded *gluteus maximus* in the human. Other muscles rotate the hips in humans, keeping the center of gravity stable during upright locomotion , as opposed to the side -to -side motion of the swaying apes. The knees , which are normally always bent to some degree in the apes are fully straightened in humans. To become this way, it involves the inward , or medial, orientation of the patella as well. This is like a torsion or twisting of the femur and tibia which reorients the knee while keeping the body facing forward, resulting in a hominid knee, or valgus knee (the knees are brought together inward under the body's center of gravity), (Olson, 1981;Aielo, *et al*, 1990). This valgus knee is the key to balance, since the bipedal gait requires the balancing on one leg while the other moves forward, the knee provides stability during motion. For this same reason, the inner condyle of the femur is larger in humans, whereas in apes , the medial, and lateral are about the same size. The human foot bears weight in three locations during a stride: the heel , ball and large toe. All three areas are greatly expanded for this reason, providing flexible structure and stronger ligaments, and the larger big toe brings into alignment then all the other toes and is now no longer opposable and capable of acting like a thumb as it does in apes, (Childe, 1939;Binford, 1968;Olson,19*81 al,* 1990).

The most important reason for such detailed study of these differences of the muscular skeletal system is for the detection of relatedness, divergence and identification of species in the fossil record. Bipedal gait and when it began to emerge can be inferred from such evidence. Habitual bipedalism has evolved several times along the way to humans, dinosaurs and kangaroos for example), (Aiello *et al* , 1990). Howver, these are only examples of convergence since the locomotor mechanics in such groups differ a great deal from one another. Bipedalims therefore is said to be a synamorphy between any extinct species and extant humans. In other words, it is thought that bipedalism has arisen only once among hominoids, and any other species possessing the trait would be considered closely related. This trait has become a defining marker within the family Hominidae, (hilde, 1939;Binford, 1968;Olson,19*81 al,* 1990).

The earliest evidence for bipedalism in the hominid fossil record is actually indirect evidence. It was actually discovered when Andrew Hill was working in conjunction with Mary

Leakey at Laetoli in Tanzania. A dried riverbed layered with volcanic ash was found with several foot depressions in it. The depressions were of a variety of animals, but also ones belonging to early hominids.The volcanic ash made dating possible, since it can be dated utilizing potassium-argon dating (technique used for rock material). They were dated to 3.7 million years, (Leakey *et al*, 1979).

The feet responsible for the prints were short and wide, with a large toe aligned with the others for the transference of weight associated with a bipedal gait. The foot proportions were found to be similar to those of the earliest know hominids. Other hominid fossils from Laetoli have come to be known as *Australopithecus afarensis,* known from jaws and teeth, and have been found to be slightly older than similar fossils from Hadar in Ethiopia. These populations formed the roots it is felt of the hominid phyletic tree. Since the earliest evidence is out of Africa, it is believed that Africa is the continent within hominoids evolved into the hominids of the Miocene. Pliocene fossil remains of australopithecines, early hominids, are found in both southern and eastern Africa and range in age from roughly 5-6 million years ago, with the mandibular fragment and teeth from Tabarin identified as *Australopithecus afarensis* dating to about 1 million years ago , falling within the Pleistocene,(Leakey *et al*, 1979; Andrews, 1987; Reader, 1989; Simons, 1989).

It should be noted that when dealing with primate fossils in general,and the hominoids or hominids in particular, given the age and the issues surrounding both contamination and preeservation, it is helpful to refer to keep in mind that the taxonomic relationships of these specimens is not always clear. Certain fossils, as the result of the often gradual or transitional nature of evolution with particular respect to the environment and surrounding populations tend to span the morphological gaps that many often attempt to try to make real whenever something new turns up in order to bring species together in geological time and establish links in relatedness. Such fossils take on specific significance though nonetheless in terms of what they tell us of the total morphological patterns evolving through time while providing the best insight still into realtedness and adaptiveness between species, and between species and their environments, (Bates, 1954;LeGros Clark, 1959; Pilbeam , 1972; Simons, 1972;Gibbons 1981;1999)

Early Speech (vocalization)

Throughout the history of mankind there have been questions as to the causes and the events that led to the evolution of speech in humans. Unfortunately, researchers have not been able to pinpoint a specific time frame or place where speech first occurred. This is because sound is not a resource that leaves physical record or evidence of existence. It is also important to note that there is a difference between sounds being vocalized or speech of some sort and language. Speech and or vocalization is the production of linguistic sounds by the vocal apparatus, while language is a symbolic form of communication. The earliest evidence of written language provides an indication that some form of spoken communication was in use. However, early written records cannot be cited as proof that speech existed. And these documents cannot indicate how long speech may have existed prior to the written language itself. Walker and Shipman, (1996) express it more eloquently when defining language: "As Petitto observed, the fact that deaf infants babble shows that language is an innate capacity in humans; it is the mode of expression, not the ability itself, that is learned. Or, to use the felicitous phrase of Steven Pinker, a linguist at MIT, there is a 'language instinct' hard-wired into the human brain", (Zuckerman, 1932;Lieberman, 1968;Linden, 1974;Johansen *et al*, 1996).

In terms of speech itself, speech is produced through coordinated movement of the organs of the human vocal tract, (vocal chords of the larynx, trachea (airway). To understand the evolution of speech, it is therefore necessary to understand the emergence and development of the neurophysiological structures that are involved incontrolling and perceiving these movements. Recent research has tended to focus either on the development of cognitive capacities associated with changes in the structure of the central nervous system, or on the morphology of the vocal tract, which imposes certain physical constraints. Less attention has been paid to the channelling of information in the peripheral nervous system, which is determined by the structure of the effectors and receptors connected to the brainstem nuclei via the cranial nerves. Peripheral feedback loops are known to be important in coordinating respiration, phonation, and oral movement to produce speech. Tongue,jaw, lip, and larynx movements are monitored by muscle spindles located within the different muscle groups, which provide information on muscle contraction, and by mechanoreceptors in the oral mucosa,which provide information on patterns of contact. The type and spatialdistribution of these receptors can be expected to play some role in determining the sensory motor representations that can be exploited at higher levels. The evolution of appropriate patterns of motor and sensory motor responses within the vocal tract may therefore constitute a probable important pre-adaptation for the control of oral movement required for the production of speech, (Lieberman *et al* 1971,1972;Gibbons, 1977, 1981; Johanson *et al*, 1996; Walker *et al*, 1996).

The basic mechanism of speech contains three main components: respiratory, phonatory, and articulatory. The role of the respiratory component in the production of sound is to produce an outward flow of air under pressure. Phonation (voicing) is produced by vocal folds that are pulled together so that they vibrate when the air flows across them (Negus 1949; Lieberman *et al* , 1972;Gibbons, 1981). The articulatory component, which is essentially the mouth, is usually opened for some time during the voicing process, and the sh ape of the cavity between the lips and vocal tract (larynx) modulates the tone and shapes the quality of the noise. In hominids, there is a two-tubed vocal tract (Lieberman 1984), which considerably increase the ability to make different sounds, an essential aspect of speech. However, of all the primates with this vocal system, only humans evolved the organization and orchestration of the various components that allowed them to be used carefully and productively for speech. Furthermore, this was perfected to allow for syllabic chunks of sound. Except for humans, mammals typically have a small range of vocal noises (calls) because of their relative disorganization of the larynx and the lack of development around Broca's area. However, it is interesting to note that the more advanced primates have a greater number of calls, and their characters are more complex (Dunbar 1975). This again lends evidence that the evolution of speech was a slow gradual process, not a case of punctuated equilibrium.

In most of the mammalia, present evidence suggests that sensory terminations appear to be concentrated along a series of transversal ridges (rugae palatinae) distributed along the length of the palate, presumed to be linked to the manipulation of food within the oral cavity. In man, however, the palatal region has undergone a number of morphological modifications, and the rugae have shifted towards the incisors. This is partly due to changes in bone structure, but may also be linked to changes in diet. It is not yet known if the spatial distribution of sensory receptors in the palate has undergone similar changes, distorting the representation of linguo-palatal contact patterns, and possibly favouring the appearance of coronals, for example, by focusing attention on the alveolar region, (Negus 1949; Lieberman *et al* , 1972; Dunbar, 1975; Gibbons, 1981).

The main physical component that separates human speech from other mammalian vocalizations is the articulatory component. Other components are identical in all mammals: the respiratory cycle of inhalation/exhalation is modified slightly to produce the desired length of the call, while the phonentic component is the vocal cords opening and closing, with different tension and pressure levels to create variations in perceived pitch. In humans, however, there is coordination between the articulatory component and the phonetic component that can produce consonants, vowels, and more elaborately, syllables. The repeated cycle of opening and closing the vocal tract is both unique to humans, and the crucial behavior that provides us with speech capabilities. In fact,virtually every utterance of every speaker of every one of the world's languages involves this process,(Lieberman *et al* , 1972; Dunbar, 1975; Gibbons, 1981; MacNeilage, 1991).

However, although this is the process that provides the critical bridge between simple primate calls and the complex nature of speech, none of it is possible without the fundamental structures shared by all primates and to a lesser extent by other mammals. These structures were not created for the purpose of speech, but rather were adopted by the speech mechanisms. It is possible to see how the evolution of the various structures and components might have been useful for other reasons, and how the y could be combined over time by looking to the primates, and it also acts as an insightful method for exploration of the vocal/speech and language abilities developed by both human and non-human primates, (Jacob 1977). Movements of the mouth in the normal mammalian activities of chewing, licking, and sucking are easily seen to be similar to common motions in speech. They also all involve "successive cycles of mandibular oscillation," which is also crucial to speech. The respiratory cycle is necessary for gas exchange, and provides "free" airflow for speech. It is thought that vocal folds were initially used to control airflow though the esophagus. The controlled, cyclic movements of mastication use the same muscles in the same manner as is necessary for speech, which lends more credence to the hypothesis that speech is closely linked to the digestive system. The proximity of the brain center for speech to Broca's area also supports this theory. Although physiologically and genetically humans are not dissimilar from other primates, psychologically as well as anatomically with respect to the hyoid bone and vocal chord funtions, they are just different enough to be able to utilize a particular range of vowel pronunciation abilites in order to form the communications which evolved along with the lineage over time, and in response to bipedal locomotion, given the morphological changes and their correlations to the eventual placement of the hyoid and larynx abilities. All the evidence points to a slow,gradual adaptation from pre-existing structures relating to respiratory and digestive systems, as well as skeletal adaptations responding to bipedalism habitually,and the transition from hominoids to hominids, (Lieberman *et al* ,1971, 1972; Dunbar, 1975; Gibbons,1977,1981; MacNeilage, 1991).

Hominids

Single Species Evolutionary Scenarios :Phylogenies and Cladistics

Phylogentic trees, which represent branching of species in terms of their ancestral divergences and therefore relatedness, are filled out (drawn) by making inferences about the selective pressures involved in adaptations and therefore speciation events, along with inferring causes of evolutionary divergences which have been traced. This information comes from such things as studies on paleoclimatology, paleoecology, or taphonomy (soil analysis). For example, one example previously discussed in this paper was the emergence of habitual bipedalism. There are a variety of explanations for its development which is the result of inferences made from available data. Not only is it recognized as a derived character, but that it was also inherited from a common ancestor (*Australopithecus afarensis),* and then attempts are made to offer explanation. Just as many different trees are compatible with a single cladogram (cladistic methodology), many different scenarios are compatible with a single tree. An evolutionary scenario will provide the maximum range of data, but it will also contain the most assumptions.

A cladogram on the other hand, will have the least amount of data and the fewest assumptions. That is why it is the first and most fundamental aspect of historical biological records. The cladogram is a branching diagram without a time dimension as in the case of phylogenies. It, too, shows the closest relatives among a group of taxa from the distribution of synapomorphies. Cladograms are constructed based also on the traits in common of species, and it recognizes that there are also two sets of traits to an organism: the original condition , primitive, and the altered one, or derived, as in the case of the tree shrew earlier in this paper. Its taxonomic status has been in question due to the conflict and comprehension of what are its primitive characteristics and its derived charaacteristics and where those put it in relation to not only other primates, but other mammals as well. Every species is a mosaic of both primitive and derived traits. Therefore, no extant species is more advanced than another, yet all have individual derived traits in order to distinguish it from others. In adding time to a cladogram , the phylogenetic tree is constructed. Descents are represented in such trees.The phylogeneitc tree therefore is a representation indicating when certain species lived and diverged from ancestral species. It contains what a cladogram contains, with the only difference being the addition of temporal data , and a good many more inferences, (Deetz, 1968;Hill et al, 1972;Tatterasall *et al*, 1977; Landau, 1984; Gibbons,1999).

Anatomical Variance

Briefly, it is important to put forth an example of interpretation. If there are two skulls found as fossils, and they are different from one another in many ways, then interpretation can be done based upon morphological traits. Morphological interpretation is one of the leading difficulties of the taxonomic process, though work being conducted today in the area of DNA interpretations is beginning to aid in such analyses.

Taxonomy and Variation

Much of the variation in the fossil record is interpreted as representing taxonomic diversity. Not only do different species look different from one another, but there are also a large number of extant species alone. So, two fossils which look different could be from different species. However, there is a tendency in the fossil record to interpret too much of the observable differences, leading to splitting, or the creation of fossils being divided up into far too many taxa. On the other hand, a tendency to interpret too little of the observable diversity results in just the opposite, lumping, or placing what may have been different species into the same taxa, (Andrews, 1978;Fleagle, *et al*, 1980;Mayr, 1988).

Sexual Dimorphism

Males and females of most of the primate species look different from one another, particularly skeletally. Males tend to be larger , have larger canine teeth, and larger sagital crests along the skull for larger muscles to attach**.** Most splitting is the result of a failure to recognize such diversity. The next issue then is how to distinguish between a male and female from the same species. Fleagle, Kay, and Simons, (1980), found that in their sample of *Aegyptopithecus zeuxis,* molar teeth were all about the same size**,** while canine teeth were either large or small. This suggested that , based on the general pattern in catarrhine primates, the ones with larger canines were the males and the ones with the smaller were females. During the 1960's some views held that large, "robust" australopithecines and the small, gracile australopithecines, who lived in Africa 1-3 million years ago might simply be males and females. Through further finds of female "robust fossils, it was suggested that lumping them would underestimate their taxonomic value, going against the single species hypothesis then by keeping them split so vastly. So, whether or not the fossils from Hadar, Ethiopia, represent a single highly dimorphic species (*Australopithecus afarensis)* or a mixture of two or more related species, large and small, has been argued over, though most settle for considering it all one species, (McHenry, 1975;Delson, 1981; Andrews, 1987).

Onotgenetic Variation

The end of the preformism-epigenesis debate in the eighteenth century (Stocking, 1987) resolved once and for all that at all stages of life, organisms do more than just grow, they also change in the process. Animals are therefore dynamic entities over time and space and possess an ontogeny. This leads to a problem though, if most of the species which have ever lived throughout time are extinct, which they are, then how is it then possible to link a juvenile form of any extinct species with an adult form? The fossil record can not account for all possibilities, and often only one species will be found from a particular group in a given place at a given time. Therefore, if an immature specimen is found (ie: the erectus skull of a child from Modjokerto , Java) they most likely belong to that species. However, if a child's skull in a context for which there are more than one hominid species , as in the case of the Plio-Pleistocene hominids of soutehrn Africa from 2-3 million years ago, how is it determined as to which species it belongs to? The criteria relied upon then becomes more subjective. In fact, the Taung skull has often been referred to as a juvenile "robust" australopithecine as opposed to where it normally is placed, among the graciles. Since it is difficult to know what these immature individuals or species looked like, then it is also just as difficult to allocate any found isolated into one species or another, (Johanson *et al* 1976, *et al,* 1979;Skelton, *et al,* 1986).

Ancestral Variation

Variation at the subspecies level is certainly common in animals. There are several subspecies of chimps for example, each very subtly yet consistently different from one another. The subspecies (*Gorilla gorilla gorilla* - the lowland gorilla, and *Gorilla gorilla beringei* - the mountain gorilla) differ in their general length of hair, size of nostrils, chest area, skull form, and various hand and feet aspects. The subspecies of orangutans (*Pongo pygmaeus pygmaeus* , the Borneo orang, and *Pongo pygmaeus abelii*, the Sumatra orang) have consistent differences in the sturcture of their faces and in their facial hair. Now then, the question is how would it be known to make them different subspecies as opposed to different species or the same subspecies if found as fossils. Again, much of this would be subjective, based upon the amount of morphological differences taken to be appropriate at the subspecies level. In one example of two subspecies of *Homo sapiens: Homo sapiens sapiens and Homo sapiens neandertalensis,* many have divided these into two separate species, while others support the subspecies decision, (Cuppy, 1931; Tinkaus, 1979; Smith *et al*, 1984;Bray, 1986).

Polymorphic Variation

In all populations , a high degree of variation is noticeable among all of the individuals who comprise them. This means that from the exact same genetic background any growing organism can develop in different ways due to given environmental pressures affecting the underlying genes. For example, two genetically identical plants can be grown at different altitudes and will most likely grown to different sizes, (Hill, 1968). Humans living in higher altitudes tend to develop certain anatomical and physiological responses to the long term stress of breathing thinner air or being in colder or warmer climates.

These are not all genetically dependent, but the traits will appear in a population only as long as the environmental stimulus is present and acting upon them. This is known as the development of ecotypes. Ecotypes are stable developmental responses by whole populations to a varied environment and its pressures when significant genetic differences are absent within the population. So, if two fossils are found which appear similar but somewhat different, it could be that these differences are due to environmental stress and developmental plasticity. Briefly, the opposite fo ecotypic variation would then be polymorphic variation. This would be the hereditary differences atrributable to shuffling of genes which occurs as a natural result of reproduction. In inter[reting these differences then between any two fossils, what then needs to be considered as well is the possibility that these differences are simply a sample of the normal genetic diversity which exists in populations, (Hill, 1968, Fleagle *et al*, 1980;Landau, 1984).

Pathological Variation

Not all fossils represent totally healthy specimens of any species, but do represent a variety of species in a variety of conditions and at a variety of ages. Some diseases in human populations such as syphylis, tuberculosis, and various vitamin deficiencies leave marks on the bones and can be studied along with traumatic injuries or birth defects. All of these create a variety of skeletal differences among individuals. This is true of any species, and among humans, cultural practices also can result in dramatic bodily changes to certain areas, especially the skull. Though not every change is a pathological variation, the results are the same, a change in morphology is seen. These sorts of variants can be detected even in the smallest sampling of skeletal material. If there are normal individuals, then the pathologies will be more spread out. The problem in this is that some specimens may be the first ones found of a given species , therefore complicating what might be considered normal as opposed to what is considered a pathological condition, (Deetz *et al*, 1968; Hill, 1968; Tattersall *et al*, 1977).

In 1913, Marcellin Boule, of France conpletely reconstructed a skeleton (Neandertal specimen) as permitting only a bent-over stooped kind of gait. This became the image of how such individuals moved on the ground. Others eventually came forth with challenges. When reanalysed by W.L. Straus (1957) showed that Boule's reconstruction was essentially correct, but that the specimen actually belonged to an older and more arthritic male, holding up that generalizations can not be made on the basis of one specimen, (Deetz *et al*, 1968; Hill, 1968; Tattersall *et al*, 1977).

Temporal Variation

Due to the fact that evolution is change through time, or successive generations, it is often found that populations of organisms living at any one time are slightly different from populations having lived at other times. Therefore, the fossils of say *Homo erectus* from Ngandon and Sangiran , both in Java, span roughly about 500,000 years. Despite the fact they are from the same locations and are similar, they differ in terms of skull size and shape. So, if two groups of remains from the same species at the same place , but from differing time periods, and show consitent differences, then extreme caution must be exercised when interpreting biological variation so as not to overly split a given speicies, (Deetz *et al*, 1968; Hill, 1968; Tattersall *et al*, 1977).

Taphonomy

Fossils are not necessarily the keys to the organism from which it came. Fossils lie in the ground for much longer than the organism ever lived, and are therefore subjected to many chemical, biological and mechanical forces: the stresses of destruction, deterioration, and deformation. Taphonomy studies what happens to the bones between the time of death of an organism and the time the remains are found. Oftentimes, fossils will be found as part of the

remains of an old den where some carnivorous organism lived. In which case, not only do the

fossils need to be interpreted from the perspective of the chemical alteration of bone into fossil

and the geological stresses of being buried and embedded in rock for millions of years, but also

in light of the fact is was subjected to the stresses of something else chewing on it with its teeth,

or by tools they may have used at the time. Bones are also continually walked over by wild

animals, and subjected to the movements of water which will determine how they wear , decay

or are preserved as well. Taphonomy there fore acts asa bridge between paleontology and

archeology, (Deetz, 1968; Hill *et al*, 1972; Tattersall *et al*, 1977).

The Australopithecines

The Pliocene

The Pliocene epoch was a rather short period, yet a time of much change in terms of

climate and fauna. Tectonic events led to the formation of such land bridges as the Panama one

joining both North and South America as well, allowing for additional faunal changes, and

variations of migration patterns between species of both continents. Temperatures also began to

warm and sea levels rose in comparison to Miocene conditions. Cercopiths began to populate the

Old World and hominids began to evolve and migrate. 10-5 million years ago though, during the

Miocene, is when the divergence between the humans and apes began to take place. Since many

pieces of the few fossils recovered though are mere fragments, then the determinations at times

between humans and apes has been difficult, (Hill, 1994. Fleagle, 1988,1999). The first known

hominids though are thought to be from the early Pliocene at about 4.4-4million years ago, (Hill,

1994). The first really recognizable hominid fossils however have been found to be from the late

Pleistocene, dating to about 2-3 million years ago, (Fleagle 1988). The Pleistocene itself was a time of great climatic changes yet again with respect to a series of glacial and interglacial periods causing greater regional adaptations to occur throughout hominid populations on a global scale, (Dart, 1925; Grine, 1988, 1993; Fleagle 1999).

Apes and Humans

The finding of the australopithecines has been one of the most stimulating and controversial continuing series of events throughout the twentieth century. They have been referred to as ape-men or man-apes , and yet seem to clearly be more closely tied to early man than to apes. The early fossil primate record, as well as almost the entire history of mammalian evolution, has indicated an emerging and distinctive pattern. The earliest groups of ancestors to mammals are anatomically quite generalized. Some of the descendents remain so, while others develop specializations. This specialization can bring about extinctions due to the fact it causes the organisms' niches to be quite narrow, and more vulnerable to change. Howver, continued generations are able to adapt over time, and therefore survive, which means they become other forms, either of the same species, or become other species eventually altogether. The other issue to consider, again, in terms of all species, and in particular respect to the australopithecines, is the idea of groups of ring species. That is , in times of favorable environmental conditions the populations are larger and breed more regularly with one another. In times of poor environmental conditions, the populations split off into smaller pockets and smaller niches. They survive, but breed only within these pockets, causing more variation to occur between them and the others. When conditions become good again, they may regroup and continue to breed, or some smaller pockets may become environmentally isolated , thus leading to so many changes over time that they develop into either a subspecies or new species altogether. This pattern is fairly evident in terms of the australopithecines, although the generalized ancestors have not positively been found as yet. The best known common ancestor to date is *Australopithecus africanus*, considered to be a small, gracile, australopithecine. Though some consider the more robust version of this to be a more climatically specialized hominid, *Australopithecus robustus* , some refer to it as *Paranthropus robustus.* Here it is now important to note though that in terms of meaningful taxonomic levels concerned with primates and primate paleontology, the meaningful level is the genus as it was evident in the consideration of the earliest fossil primates (the species is to be sunk back to the earliest known genus of its kind), (Simons *et al*, 1965, Simons, 1972, Delson, 1985; Fleagle, 1988, Grine, 1988;White *et al*, 1991; Wolpoff, 1996).

Australopithecus africanus

In 1924 raymond Dart, then an anatomist as the University of Witwatersrand in Sout Africa found the first fossils which were more man-like than ape-like. In 1925 Dart presented a partial skull and brain case of this first australopithecine, or southern ape from which the name derives. Since no extant apes live in this part of Africa, it was significant even if it had been more ape-like. The skull was found at a limestone quarry near the Taung railway station in Botswana and is thought to be of a child, hence the name "Taung Child", (Dart, 1925). The first premolar teeth had been erupted placing the age of it at roughly 6years old in terms of what was known of dates of dental eruption at the time, though it is hard to know precisely what the rates of growth and maturation were during the time this specimen lived. Also, though it is not common to define a fossil on the basis of only one immature specimen, Dart and those who followed in reanalysis, found that this Taung specimen exhibited many important diagnostic traits found among the australopithecines overall, particularly in terms of the hominid traits of the skull. Though the australopithecines were for the most part originally viewed as some type of transitional ape, especially since upon first glance the skulls of both the perceived graciles and robusts appeared more pongid than hominid, but the overall pattern has been found to clearly indicate hominid, (Dart, 1925;McHenry, 1975;Howell, 1978; Fleagle, 1988, 1999; Grine, 1986; Wood, 1991).

Dart's adversary was Robert Broom, a Scottish paleontologist and physician. Between 1936 and 1948, Broom collected adult specimens of both the gracile and robust specimens from limestone caves. The most important thing to turn up in his discoveries was an adult australopithecine believed to be female, found at the site known as Sterkfontein. It was a complete skull, with only a lower jaw missing, and it confirmed the hominid pattern found in the Taung Child. However, being adversairous in paleoanthropology has always seemed to mean that each new find should be given a new designation in with respect to genus and species. So, Broom designated his specimen *Plesianthropus transvallensis.* It has since been reclassified as *A. africanus.* Additional australopithecine material was found at Makapansgat, about 150 miles north of Johannesburg. Broom found two more sites then in the Transvaal , here considered to be the first of the robust material. Due to Broom's work, in conjunction with J.T. Robinson, there emerged samples of both males and females as well as general age groups of the southern African australopithecines, (McHenry, 1975;Howell, 1978; Fleagle, 1988).

Dating had been difficult throughout time, given that chronometric dating could not be applied to the formations, and dating by faunal remains has only made it more complex. It is believed that the graciles appeared earlier on than the robusts, and dating at sites such as Sdterkfontein and Makapansgat which are gracile sites are older , dating some two to three million years, making the australopithecines individuals of the late Pliocene and early Pleistocene, (1925;McHenry, 1975;Howell, 1978; Fleagle, 1988).

The *africanus* Pattern

The gracile australopithecines possess a total morphological pattern that is clearly hominid. This is based upon a number of cranial features primarily. The skulls of *Australopithecus* shows a well developed forehead. Some frontal lobe brain expansion was therefore likely to have occurred. The supraorbital region is not well developed however, as there are no supraorbital ridges or a barlike type torus. There is also a more elevated and more developed cranial vault (braincase), and the main areas of the brain were placed higher than the level of the face, not below it and more towards the rear as in the pongids. The cranial capacity range falls in at about 450-600cc, which is considered more of an ape-size though. However, the brain size is not a clear cut indicator of overall development. With respect to overall body size, these individuals were roughly four feet tall and their maximum weights probably were about 80 pounds. Though the australopithecines had a larger brain than a great ape, it was more of a pongid size. The nuchal muscles wer not all that heavy as they are in pongids, as is evident by the lower placed occipital torus on the rear of the skull. This bone is a bar which runs across the back of the skull and marks the height , and to some extent , the nature of the neck muscles. In the case of *africanus*, the torus is not all that robust. Given this morphological data, it is evident that these individuals possessed and erect posture, since heavy neck muscles are needed in order to support a larger skull, and particularly one which was not all that well balanced as yet on top of a spine. Additionally, an erect posture is also supported by the presence of forwardly placed occipital condyles and foramen magnum. Australopithecines had large jaws, giving the skull a somewhat pongid appearance, in addition to the extreme prognathisism which also was present. This protusion of the jaw was most likely due to the fact that these gracile individuals had larger incisor and canine teeth which requires a good amount of butressing by bone. These jaws supported a definitve hominid dentition. The teeth were arranged in a parabolic arcade coming closer to the appearance of the anatomically modern human dental arcade, and in terms of the size of the anterior teeth, they have been found to be proportionate to the posterior teeth (premolars and molars) and this is said to represent what is referred to as a balanced dentition. While the dentition was large, the muscles for mastication that supported it were not all that large themselves. There was no sagittal crest needed for anchoring of the temporal muscles. The masseter, another prime muscle of mastication, was also not all that large, therefore, the zygomatic region (cheek) was not that robust as it did not need to absorb a great deal of chewing force, (Dart, 1925;McHenry, 1975;Howell, 1978; Fleagle, 1988, 1999; Grine, 1986; Wood, 1991).

So, given what is known of both the cranial and postcranial remains of these australopithecines, they were indeed erect and bipedal hominids. The iliac portion of the pelvis presents as more broad and short, another key feature in a bipedal hominid. Therefore, the total morphological pattern is one which supports Dart's (1925) conclusions early on , as well as the knowledge of what an early anatomically modern human ancestor looked like.

Robust Australopithecines

Although not all that much more different in terms of appearances or in comparison say to the more robust appearance of later classic Neandertals, Robert Broom also had found that there seemed to be enough anatomical differences to decide to place these particular australopithecines as a more robust version of the graciles. Broom originally named then in their own genus, *Paranthropus*, though they are acknowledged to be australopithecines and are more appropriately placed back with *Australopithecus.* These individuals also lived in South Africa around the middle of the Pleistocene, about a million or so more years later than *africanus* did. They are known from the Transvall site, and in 1938 were found at Kromdraii and again in 1948 at Swartkrans.   What is striking to note about these so-called more robust australopithecines, is that the Mary Leakey find in 1959 of a skull, then named by her as a new genus and species (*Zinjanthropus boisei)* is now known to be merely a variety of the already known East African robustus. What is important about this find though is that it is evidence of the fact that this australopithecine did in fact inhabit other regions of Africa and was in fact variable due to regional adaptations, therefore explaining the slight differences between these robusts and graciles. Most of what the Leakey's found again even in 1968 and 1969 is also similar material , supporting the wide range these individuals inhabited along with the existence of both sexual dimorphisms and regional adaptations in morphologies.   Some robust material is also known from Java in China, again supporting a wide range of distribution and therefore variation within the genus,(Howell, 1978; Fleagle, 1988, 1999; Grine, 1986; Wood, 1991).

The Robust Pattern

The total morphological pattern of these said to be more robust australopithecines is best described as that of an aberrant hominid, and as is the case with the more gracile australopithecines, these also have pongid features though they are quite hominid. This particular group happened to adapt to an econiche that was either similar to a gorilla's or at least required some gorilla-like features for survival. Though these particular australopithecines are considered to have been a lineage which veered off the main hominid evolutionary path leading to a dead end, there are some indications that they may have been slightly more robust due only to regional variations which developed , and that their traits could have been found in other populations to the extent that they contributed to future gene pools which possessed genes capable of being expressed under certain environmental conditions, for example as a case of preexisting variation which could have been expressed in the much more robust features of the classic Neandertals, though this is only one thought. These individuals were also about four feet in height , but their weight ranged up to roughly 120 pounds, though they had similar (450-600cc) cranial capacities as the others, all adding up to the fact that these were also australopithecines which did not differ much from what was thought to be the more gracile version, other than the fact they displayed some regional variation, (Howell, 1978; White, 1981; Lovejoy, 1981; Fleagle, 1988, 1999; Grine, 1986,1988; Wood, 1991).

The frontal region of the skull had a slightly larger supraorbital ridge in comparison back to the graciles, though the forehead is barely present, and the cranial vault was not found to be as rounded or as expanded. A slight sagittal crest was also evident, at least in the males, along with a larger zygomatic region and jaw. The occipital torus was slightly more raised, indicating that the neck muscles were slightly more massive than those of the graciles. The foramen magnum and occipital condyles were also slightly more forward in robustus, and probably related to the jaws not being all that prognathic. This decrease in prognathicism is due to the fact that the anterior teeth were smaller and the molars larger, an unbalanced dentition, yet the parabolic arcade remained. The larger molars, and decreased prognathicism were most likely adaptations to a more crushing and grinding type diet, (Fleagle, 1988, 1999; Grine, 1986,1988; Wood, 1991).

What is known of the postcranial anatomy seems to indicate a mixture of pongid and hominid traits yet again. The hand bones indicate a good possibility that these individuals may have also been knucklewalkers to some extent, along with some features of the pelvis. The pelvic remains from Swartkrans for example, show differences of a more elongated pelvis with some curvature, supporting the idea that they may not have been completely adept at bipedalism just yet, though a better biped than either chimps or gorillas,(Fleagle, 1988, 1999; Grine, 1986,1988;

Wood, 1991).

Controversies:

1. *Homo habilis*

From 1925 to the present, the australopithecines have been the primary center of continued debate and speculation. In terms of What 1964 brought in the redefining of the *Homo* genus in order for Leakey, Tobias and Napier to pull off a naming of an entirely new species based on juvenille remains found in Olduvai Gorge, claiming it to be a more advanced australopithecine, though it more closely resembles *africanus* morphologically, and most feel it is an Eastern African version of this australopithecine,(Tobias 1967;Johanson *et al*, 1979; Fleagle, 1999).

2. How Many Genera:

Paranthropus robustus V. *Australopithecus robusts* and *A. africanus*

One of the major australopithecine debates has been over the issue of whether the gracile and robust australopithecines are two biologically distinct genera or if they are merely two species within the same genus. According to J.T. Robinson, (1972), is the prime adherent to the two genera school of thought. During the 1960's he devised what has become known now as the dietary hypothesis. He claimed that the differences in the dentitions of the two astralopithecine types represented major dietary differences. The gracile form was more of an omnivore, and thus a hunter-gatherer. The larger canine teeth he said were a carnivorous adaptation. In the case of the robust types, he claimed an all vegetarian diet which included seeds, roots, and bulbs. This type of vegetation has sand or silica woven into it resulting in excessive wear patterns and scarring seen on the molars. Robinson based his hypothesis partly on evidence that the gracile australopithecines lived in rather arid times when vegetation would have been less abundant and animal protein a more reliable source of food, while the robusts lived in wetter times when vegetation was more abundant. The only thing this hypothesis really supports is that these were contemporary species who's diverse econiches accounted for specific regional adaptations leading to their minor morphological differences, (Robinson, 1972; Grine, 1981, 1986, 1988, 1993; Fleagle 1999).

One argument favoring Robinson's position though has been the Competitive exclusion Principle. This hypothesis holds that two species of the same genus could not both survive for long in an environment in which they were competing for the same econiche. One would either become extinct or evolve into another niche, likely to become a new species and or genus. Similar creatures can live in the same environment and seem to occupy the same econiche but actually be adapted to less obvious microniches that would not be apparent in the fossil record, and example being diurnal and nocturnal niches. Therefore, it is more appropriate to classify these two forms as two species within one genus, *Australopithecus*, as both regional adaptations in conjunction with sexual dimorphisms are the most likely cause of morphological variance, (Hutchinson, 1965; Gibbons, 1981; Grine, 1988; Fleagle, 1999).

Homo erectus

A more advanced hominid had begun to emerge during the middle part of the Pleistocene, roughly 2.5 million years ago, which was the first true representative of the genus *Homo*, (Rightmire, 1990). This Asian species is similar to the African version , or *ergaster* by some, giving it a separate species status, (Wood, 1992; Fleagle, 1999). Wood , (1992) has felt *ergaster* displays distinct cranial proportions in comparison. However, other analyses between the African and Asian specimens have yet to find consistent and considerable differences, leading them to both fall under one species, *erectus*, (Gibbons, 1981;Rightmire, 1992; Walker *et al*, 1993; Brauer, 1994).

Traditionally, it has been felt that the African remains were older than the Asian ones, with dates having ben reported early Pleistocene dates for the appearance of early *Homo* in both China and Indonesia, (Swisher *et al*, 1994). If so, then *Homo erectus* was established on both continents from the early Pleistocene right up through the middle Pleistocene, (Ciochon, 1996). To date, no fossils exist which can as yet place the genus and species in Europe. However, the geographic distribution of this genus, and of *erectus* in particular, is far wider than of any other species prior to this time, (going back to it's initial appearance of roughly 2.5 million years ago). This temporal and spacial span continue to be issues of debate. In reexaminations conducted during the late 1990's of fossils from the Solo River in Java have indicated though that the *erectus* species survived at least until about 27,000 years ago in Southeast Asia, (Swisher *et al*, 1996).

In comparison to the australopithecines, and even *H. habilis* (going back to the issue of how Leakey, Tobias and Napier redefined for themselves the genus in order to place their find within it due to its tool traditions as previously discussed above) , H. *erectus* has smaller cheek teeth and a thinner mandibule, supporting a reduction in dentition within the genus. It's brain size is larger than in the earlier hominids, though it is important to note that considerable amounts of variation have been seen throughout the species, (Brown *et al*, 1985; Fleagle , 1999). The cranium is characterized by thicker bones, a longer and lower placed vault with some sagittal cresting , projecting brow ridges and a more prominent occipital torus. The face is more broad and the nasal passages are larger, with an overall body size a bit larger than the australopithecines but with less sexual dimorphism present, (Rightmire, 1992; Walker *et al*, 1993; Brauer, 1994).

The most complete specimen thus far known is from the lake Turkana in northern Kenya and dated on the basis of deposits in the area to about 1.6 million years ago, (Brown *et al*, 1985; Walker *et al*, 1994). It is of an approximately 12 year old male who stood an estimated six feet in height, with similar limb proportions to those of the later *H. sapiens*, but with a more robust appearance and a chest which was found to be more conical in shape, or more ape-like still, along with a femoral neck which was longer as is found among the australopithecines, and yet a larger femoral head characteristic of anatomically modern *H. sapiens*. Overall, it was seen as being similar to specimens found in northern latitude populations, most likely this specimen represents mosaic evolution and a transitional phase in hominid evolution,(Fleagle, 1988, 1999; Ruff, 1993).

Homo sapiens

Antomically modern *H. sapiens*, similar to living populations of our species, begin to emerge in the fossil record from Africa as well as the Middle East about 100,000 years ago and about 40,000 years ago in Europe, with what are intermediate forms showing up even prior to this, (Brauer *et al*, 1997). This species is best distinguished from previous ones by its smaller teeth, vertical mandibular ramus, chin formation , much more shortened face overall, very little pronouncement of brow ridges, and a higher vertical forehead. Even in a comparison to Neandertal specimens the limbs are thinner and the trunk longer and more narrow, a definite indication of climate adaptive strategies affecting both types of populations of individuals, as the markedly lower robustness of the overall skeleton of *H. sapiens*, and subsequently of the early *H.*

sapiens sapiens (anatomically moderns), during the later Pleistocene. This species far surpassed the geographic ranges of hominids coming before them, and successfully colonized many varied climatic regions of the North and South America, Asia, and Europe, (Gibbons, 1981;Rightmire, 1992; Walker *et al*, 1993; Brauer, 1994).

Phylogeny

The main issues are concerned with whether the various populations and species of fossil humans (hominids) are part of a single lineage (single species hypothesis) or whether there have been multiple lineages at any one time with some giving rise to others while some became extinct. This has been an issue for every transitional (intermediate) specimen found of hominids dating back to roughly the past 2 million years. The fossil record has been limited though, by issues of poor preservation, small pockets of populations leaving no evidence of having lived, issues of geographic adaptations and sexual dimorphisms. Therefore, phyletic details cannot be clearly resolved either purely morphologically or purely through DNA analysis, but is best done when such techniques are combined making inferences more reliable. On this basis then, there were most likely only three species of australopithecines (*afarensis, africanus, robustus*) and at least two species of *Homo (erectus and sapiens)* , as well as two subspecies (*sapiens sapiens and neandertalensis)*, living in and subsequently arising from Africa. Though questions still remain in terms of seperating the australopithecines as previously mentioned into two genera, as well as splitting the *erectus* material from Africa and the *ergaster* material from Asia, and the ongoing issues over the placement of Neandertals. The simplest and most likely scenario is the single species hypothesis, three species of australopithecines, two species of *Homo* and it's two subspecies. In terms of a single species, though not all would describe hominid evolution from this perspective as an adaptive radiation, this is a lineage which shows close relatedness yet still developed remarkable adaptive abilites in order to inhabit such a wide geographic range, (Gibbons, 1981; Thorne *et al*, 1981;Wolpoff, 1994). Further down a more detailed discussion regarding the various models of hominid emergence and relatedness follows, as well as analysis pertaining to the issues of morphology and molecular techniques.

The Neandertals

It was during the Pleistocene epoch (2.5mya - 11,000years ago),(Binford 1985; Melentes *et al* 1989), that the genus *Homo* began to emerge, and by approximately 130,000-13,000ya, *Homo sapiens* began to appear in the fossil record. It was during this period, the Upper Pleistocene, that early *Homo sapiens* began to be supplemented by anatomically modern *Homo sapiens*, (Trinkaus *et al* 1992; Treisman 1995). It is at this point that the controversy in the fossil record of modern man is at its most dense, particularly in Western Europe. Alongside anatomically modern *Homo sapiens* in Europe were the Neandertals, found in the range from the Atlantic seaboard to Uzbekistan and south to the Middle East and possibly parts of northern Africa. They are known in this region from 120,000-35,000ya, and characteristic Neandertal features have been seen in earlier African *H. sapiens* specimens. The seemingly rapid disappearance of Neandertals is based upon fossil and geographical evidence, (Butzer *et al* 1975; Treisman 1995). They disappeared quickly from Europe, giving rise to the popular myths that they were somehow exterminated by, or perhaps even interbred with, *Homo sapiens,*(Cavalli-Sforza *et al* 1988; Treisman 1995; Smith *et al* 1997).

The modern human origins debate within paleoanthropology has become polarized between two dominant models, Recent African Origins (RAO) and Multiregional Evolution (MRE). The debate has persisted and shows no real sign of resolution despite the incorporation of new data over the past decade. The RAO model holds that moderns emerged out of Africa relatively recently, probably over the last two or three hundred thousand years, (Wilson and Caan 1992; Frayer et al 1993). The opposing view (MRE) is that modern man has resulted from parallel evolution in different regions, producing convergent modernization of local populations over the last million years or so (Frayer et al 1993; Roberts 1999). Proponents of both views believe that their interpretations are irreconcilable despite the introduction in recent years of mitochondrial DNA (mtDNA) evidence, which offers a basis for yet two additional models of human evolution that are compatible with the evidence of all views. These models would propose mechanisms by which complete replacement of archaic mtDNA may have occurred in a population by recent admixture of archaic and modern types, supporting a hypothesis that Neandertals and *H. sapiens* did in fact interbreed, yet referring to Neandertal populations as hybrids, with one model accounting for migration of African populations and one not, (Wallace 1992; Wilson and Caan 1992; Templeton 1993; Treisman 1995).

Anatomical differences found to exist between the various specimens of hominids collected has added fuel to the debate, however, since genetic evidence tends to lend support to the hypothesis that Neandertals and *H. sapiens* interbred, then the noted anatomical variety may be attributed to geographic adaptation by given populations. Most likely such evidence will eventually demonstrate that only one species of hominid existed. It is likely that adaptation of various populations occurred, leaving Neandertals to be classified as a subspecies of *Homo sapiens*. The cold periods of the Pleistocene epoch directly caused changes in hominid evolution by forcing populations to adapt to extreme cycles of cold, as this period is the best-known glacial period of the earth's history. Ice sheets at one time covered large parts of Europe, North America and South America, as well as small areas in Asia. The ice sheets of Europe radiated from Scandinavia, Finland and NW Russia to Northern Germany and the British Isles. The glaciations of the Pleistocene were not continuous, but consisted of a number of glacial advances interrupted by interglacial stages, during which the ice retreated and a comparatively mild climate prevailed, (Jorde 1991; Chang *et al* 1993; Roberts 1997). Both Bergmann's and Allen's rules support climatic adaptation by a population, as both pertain to body type and limb size being a direct result of environmental factors. Bergmann's states that the mean body weight (increased mass) of a population of a species increases with a decrease in average environmental temperatures, with the reverse also holding true. Allen's rule is related to limb size, stating that as temperatures decrease shorter and thicker limbs will result in relation to the increase in body mass. Also, Beal's hypothesis, (Wolpoff *et al,* 1997), pertaining to head form and climatic stress can also be seen as support for population adaptation and variances appearing in fossil evidence between various geographic locations of discovery. This states that in colder environments that a rounder head has less surface area to body mass and is adaptive to cold. Golger's rule, (Wolpoff *et al,* 1997), brings this into focus by stating that the phenotypic differences seen among populations of the same species are merely the result of clinal variation, or adaptation to a specific geographic region (Leibermen 1982; Kollett *et al* 1993; Harold *et al* 1997; Park 1998).

Since human evolution consists of chronological changes in gene regulation, a process activated by hormones, then consideration must be given to a variety of both exogenous and endogenous factors intermittently affecting this process. Also, since hormonal levels would regulate many physical adaptations, then it may be that periodic variations in gonadal androgen, testosterone, and dehydroepiandrosterone (DHEA) played a significant, yet overlooked, role in hominid evolution. The cold periods of the Pleistocene directly caused changes in hominid evolution by causing selection for individuals that produced more DHEA (increase provides an advantage during the cold), thereby increasing body weight, bone density, and impairing fat synthesis to ensure more efficient thermogenesis. Individuals producing more DHEA were able to produce more heat from the same nutrition. As cold decreased available sustinance, individuals could derive more benefit from sparse nutrition thus giving them a survival advantage (Tagliaferro *et al* 1986; Bobyleva *et al* 1993; Baulieu 1999;Howard 1999, 2000).

The first significant cold of the Pleistocene could have resulted in increased amounts of DHEA being selected for that simultaneously produced a more robust body confirmation and the larger cranial capacity found among the remains of Neandertal populations which occupied territories with subarctic climates. DHEA also affects total serum alkaline phosphatase, the enzymatic marker of bone formation, and possibly a key to explaining anatomical differences and/or similarities found among the fossil remains of both Neandertals and *H. sapiens* in the region. It may also provide an explanation for the nature of the fossil evidence in terms of supporting the hypothesis that the anatomical differences are the result of biological adaptations to a given environment, thus offering clarity on the place of Neandertals in hominid evolution. It may also help explain what became of their populations. The climatic cycles of the Pleistocene could have had a major impact on the course of hominid evolution that has yet to be fully explored at the physiological level.

Pleistocene's Climatic Setting and *Homo sapiens neandertalensis*

Those populations most likely ancestral to *H.s. neandertalensis* are documented in the Middle Pleistocene, that period post-dating the Brunhes-Matuyama paleomagnetic boundary at about 0.7mya. (Butzer *et al* 1975; Rightmire1990;Graves1991). Stringer *et al* (1979) classified several later European *Homo sapiens* as intermediate specimens dating between the Middle to early Late Pleistocene, intermediate between what is known as *erectus* and what is known as modern. This was due to a commonality of anatomical characteristics. Many of these European specimens are thought by some to display similarities in anatomical features to many Asian specimens as well, and are sometimes considered to be "pre-neandertals", (Stringer *et al* 1979; Fleming 1995;Rogers 1998). Such observations would serve to imply that there were most likely several smaller populations in Africa of one species, all displaying degrees of regional, or local, variability phenotypically. It would also lend support to the idea of a series of migrations at various points in time of such populations, as well as to the hypothesis that these intermediate populations did in fact have genetic contact which not only affected allele frequencies, but overall geographic adaptation. The above observations also imply that perhaps at some point, due to geographic or cultural factors, some intermediate populations became isolated.

Major fluctuations in global climate which occurred over the past three million years have been associated with a periodic build-up of the ice cap over the northern hemisphere, (Shakelton *et al* 1977; Edlund *et al* 1995; Roberts 1999). A continuous record of these climatic fluctuations has been preserved in marine sediments, where it is manifested as variations in microfaunal species. However, it is the oxygen isotope record from deep-sea sediment cores, which is considered to be the best indicator of the climate of the Pleistocene, as it can be an indication of the volume of ice deposited on the land, (Shakelton 1977, Bloom *et al* 1979; Coop 1989; Hayes *et al* 1991).

The ocean water vapor is enriched in the lighter (16) O isotope. During the glacial periods (16) O isotope would be concentrated in the ice on the land by precipitation of the atmospheric water. The ocean water, depleted of (16)O, would therefore be rich in (18) O, the heavier, isotope. The fossil record of (18) O/(16) O in the seawater during the glacial and interglacial periods is preserved in the marine sediments. Such a ratio provides an indication of increased ice volume, the stages of ice increase, minimum ice periods, and information for comparison to present day conditions. The isotopic ratio accounts, though, for only about 10 percent of the time over the past one million years, (Bloom *et al* 1979; Coop 1989; Hayes *et al* 1991;Bonatti *et al* 1994).

At the height of the Weichselian glaciation, mean annual temperatures in Europe were 12-14 C lower than present temperatures, and at approximately one million years ago Western Europe was cold and dry. Cycles of colder or more humid periods correlate directly related to the isotopic ratio. Also, Sarnthein (1978) once demonstrated a correlation between the marked increase in area of sand dunes and deserts in Africa during peak glaciation periods demonstrating fairly diverse geographic conditions existed in areas inhabited by hominid populations, (Sarenthein 1978; Coop 1989; Hayes *et al* 1991; Wiley 1998). Most Neandertal remains have been found throughout parts of Europe where the conditions have been recorded to be cold and dry, and eventually somewhat isolating. However, it is also important to note that remains of erecti have also been discovered in parts of Europe, as well as in other parts of the tropical old world. They have been dated from about 1.6mya to 250,000ya, (Trinkaus 1986; Wolpoff *et al* 1986;Edlund *et al* 1995; Fleming 1995). This information supports the hypothesis of interbreeding among the various populations of what has been known as *erectus* and what has been known as Neandertal. This rules out the idea that Neandertal populations were completely isolated. A non-glaciated and non-forested area along the Mediterranean littoral probably accounts for Neandertals populating the Iberian Peninsula past the glaciated Pyrenees, and for other populations, which were found to have inhabited the Middle East and southern areas of Russia. The corridor from Europe to Asia was also more widely opened by retreat of the ice sheet during warming periods, thereby leading to limited gene flow between the established

Neandertal populations and a variety of other intermediate *Homo* populations (early archaics),

(Hayes *et al* 1991; Wiley 1998). This offers explanation to the scattered classification of several

"types" of Neandertals as well, while simultaneously supporting the blending seen between

Neandertals and the genus *Homo* specimens. Basically, this supports, again, the hypothesis of the

existence of one species with a variety of regional (clinal) intermediate forms. It also

demonstrates variety in allele frequencies for similar adaptations to a given geographic region.

Another factor to consider concerning biological adaptation would be the cultural changes that

may have brought them on.

Population movements out of Africa were probably driven by the onset of dry, arid conditions

in local regions leading to the necessity of seeking new food sources. Another possible

contributing factor may have been an increase in population density leading to increased

competition, and greater pressure to search for new food sources. All things considered, the

changing climate of the now desert regions led to marked floral and faunal changes in these areas

thus affecting both food supplies and food choices. The adverse effects of these changes on the

existing subsistence strategies, access to more and more limited food sources, and population

pressure of hunting and gathering hominids in North Africa and Southwest Asia played a role in

changing the migratory patterns of individual populations of hominids. Harsh, arid conditions of

South Africa may have been the most significant of the climatic driving forces, leading to

population movement into the cold, but more productive regions of Europe in terms of

subsistence. This migratory pattern led to the need to develop new tool traditions in order to

develop new subsistence systems, leading to the biological adaptations necessary to manage a

new diet, climate, and social structure, (Day 1969; Smith 1976;Shakelton 1977;Stringer *et al*

1979;Wolpoff *et al* 1986;Trinkaus *et al* 1989;Edlund *et al* 1995; Fleming 1995; Wily 1998).

During the early part of the Middle Pleistocene, allele frequencies changed as a result of selection pressures in various geographic regions, leading to the continued adaptation of intermediate members of the genus *Homo*. The hominids, which became differentiated and noted for inhabiting much of Europe eventually came to be what is now known as Neandertal. Dates for many of the remains of specimens are approximate, as even the oldest (Mauer) may be less than 450,000BP. Although dating has been imprecise, it has been established that the fossils cover a substantial span of the Middle Pleistocene. Coon (1991) and Wolpoff (1993) have argued that this change was continuous in different geographical regions. They have assumed that the change was gradual and those populations of late *erectus* were succeeded by populations of *sapiens* in unbroken progression. Neandertals then constitute intermediate populations within the same species. Few groups can be considered to be extinct under these circumstances,(Tattersall 1968; Trinkaus 1986;Wolpoff *et al* 1986; Rightmire 1990,1995; Fleming 1995; Holliday 1997; Park 1999). Then, since no other fossil candidate exists, anatomically modern humans can be said to have evolved from such archaic populations.

Since archaic *Homo sapiens* have been found in Africa, Europe and Asia, the point of origin must be one of these locations. Most argue that diffusion occurred out of Africa. Cavalli-Sforza *et al* (1988) have indicated that moderns diffused out of Africa at about 90,000BP, using the discoveries at Border Cave and the Klassies River Mouth Cave to support these findings. Border Cave fossils have been controversial, however. The findings from the Klassies River site had been shown to be conclusive in dating anatomically modern humans there to about 100,000BP (Fleming 1995; Vigilant *et al* 1997). Since it has become generally accepted that the point of origin was Africa, and that moderns did evolve from archaic humans, then this would intimately bind archaics, Neandertals and moderns into one species. There have been a number of questions surrounding the classification of Neandertals. Should they be a species or a subspecies, (included in *Homo sapiens* as a geographic variant or intermediate form)? Another point of consideration would be the date controversy between *erectus* and modern. It is believed by some that *erectus* is as recent as 250,000ya, with some early modern fossil evidence emerging as early as 300,000ya. If so, then this also lends some support to the idea of one species diverging into two populations, adapting to specific geographical conditions, because Neandertals appear about the time of migration of *erectus*, (Gould 1992; Brauer *et al* 1997; Holliday 1997;Park 1999).

It is important to consider Neandertal morphology in terms of its apparent theromoregulation of body heat in relation to colder environments. Trinkaus (1989) mentions: low crural and brachial indices, heavy muscle attachment sites as reflected in large epiphyses, robusticity of diaphyses of limb bones, enormous chest cavities, and the elongation of the pubic bone in the pelvis. These traits, the result of adaptation to colder environments, have been found to be present in neonates, and young individuals both at Amud Cave and Regourdou 1 Neandertal sites (Trinkaus 1989; Rak *et al* 1994; Vandermeersch *et al* 1995). Neandertal remains have been found in Europe, Asia and Africa dating between 125,000ya to about 35,000ya, a time span that is comparable to that of the migration dates out of Africa (92,000ya -125,000ya). So, with this in mind, it is likely that several populations of erectus did exist at one time becoming differentiated both anatomically as well as behaviorally and due to adaptations necessary for survival in their respective geographic regions. There are ongoing debates regarding the origins and relations of Neandertals to moderns, and conflict exists even within the genetic realm. However, the areas of variation and adaptation by populations may offer clarification, as periods of isolation and migration of old world hominids may have been related to cycles of cold and warmth every 100,000 years or so during the last one million years, (Wolpoff *et al* 1986; Rak *et al* 1994; Park 1999).

Variation and Adaptation

Anthropology has long been concerned with human social evolution. Some have viewed the process as entirely constrained by biology, others as independent of it. Studies over the past two decades began the use of social evolution as a means for solving biological problems, particularly the transition to anatomically modern *Homo sapiens*, (Binford 1981; Gargett 1989; Gilman 1983; Halverson 1987; Mellars 1988; Johnson 1989; Trinkaus *et al* 1989). Evidence for the hypothesis that Neandertals were not related to moderns has been mounting and now also consists of studies by Lieberman (1984,1989), Lieberman and Crelin (1971), and Lieberman, Crelin and Klatt (1972) of the Neandertal vocal tract which suggests that Neandertals were incapable of fully human linguistic communication, studies of mitochondrial DNA (mtDNA) that identify what is referred to as a post-Neandertal common ancestor in South Africa some 200,000ya, (Gould 1987; Mellars *et al* 1987; Stringer *et al,*1988) along with supporting geological and archeological evidence, (Gould 1987; Mellars 1988). The basic question must be why migrate in the first place? Migration is normally seen as a response to an increase in population size resulting in stress on resources and increased competition, becoming therefore a social response, (Johnson 1989). It could also be a response to climate change. This supports the idea of *erectus* becoming several populations with one or more migrating early on, at about 1.5-1.2 million years ago, (Wolpoff *et al* 1986; Rak *et al* 1994; Park 1999).

Climatic evidence, as previously stated, supports the existence of the coldest temperatures of the Pleistocene correlating with the appearance of the "classic" or European neandertal populations around 150,000ya. This cold period also could have been brought about in part by the eruption of Toba, Sumatra, the largest known volcanic eruption of the Quarternary. The resulting volcanic winter could have eradicated some populations while others migrated in order to survive, eventually adapting locally. This volcanic winter could have reduced populations to levels low enough for local adaptations and rapid population differentiation to become part of the gene pool. During this time, it is possible that some interbreeding did occur, perhaps selectively and out of necessity, but then also isolation and geographic differentiation increased. This could have led to the emergence of Neandertals, and during the warming trends to follow to moderns as well, (Ambrose 1998;Hawks *et al* 2000). It is apparent that crucial aspects of hominid evolution and dispersal relate to global climatic trends, particularly with respect to cooling and drying associated with the northern hemisphere. These climatic trends are what prompted the existing hominids to adapt behaviorally and biologically. However, it would not seem as though both the genetic and fossil evidence has accounted completely for this adaptation, let alone degrees of variation.

Published literature now includes data from several thousand samples of mitochondrial DNA, all so similar that they might have been from a population of only a few thousand individuals. Perhaps best explained as a speciation event-giving rise to not only hominids, but also specifically to the genus *Homo*, thereby establishing one species comprised of many populations. Eventually some would choose, or be forced, to migrate thereby establishing residency, starting over the course of one million years ago throughout the Old World. Over time, selection pressures dictated allele frequencies towards local adaptive needs thereby accounting for fluctuation in variation in both the fossil and genetic records, or depending upon perspective, lack of high rates of genetic diversity, (Brown 1980;Frayer *et al* 1993; Wolpoff 1989; Wolpoff *et al* 1988; Caan *et al* 1997; Balliard *et al* 1998).

It is important to note the degree of genetic differentiation among populations, particularly among lineages within the same genus. Populations belonging to a species are capable of change independent of other populations within the same species. More than a few studies have documented isolation, reproductive or other, limiting or precluding gene flow among different populations that could justifiably be called a species, but that show very little or no observable genetic differentiation based upon a variety of DNA data, (Smith *et al* 1990;Ruvolo *et al* 1993; Balliard 1997).

Anatomical evidence supports the idea that adaptation to a colder environment occurred. Simply put, pre-existing allelic variation could have been enough to propel population adaptations as a result of certain selection pressures. It can be seen in such traits as brow ridges, occipital buns, and pelvic morphologies that similar traits already may have existed to some extent at the point of origin population, thereby also providing for an explanation for the so-called non-European Neandertal. The potential for change existed, and given the climatic conditions, allelic frequencies enhancing or favoring certain adaptive traits were increased over time, and continued over successive generations altering in frequency in response to both environmental and social conditions. Populations of *Homo erectus* spread out of Africa shortly after their emergence, as fossil remains of both *erectus* and *neandertalensis* are known from Indonesia (Java), China, and the Middle East, as well as Africa, (Wolpoff 1986, *et al* 1989; Trinkaus 1986;Aiello 1993; Escalante *et al* 1994;Battisti *et al* 1996).

Trinkaus (1989) considers Neandertal morphologies in terms of their apparent thermoregulation of body heat in relation to cold environments. Low crural and brachial indices, heavy muscle attachment sites as reflected by large epiphyses, robusticity of diaphyses of limb bones, enormous chest cavities, and elongation of the pubic bone of the pelvis are seen as functionally suited to cold. These traits are considered heritable and adaptable as they are also found to be evident in neonates, and young individuals both at Amud cave and the Regourdou 1 Neandertal sites (Rak, Kimbel and Hovers 1994; Vandermeersch and Trinkaus 1995). Also, the cranium displays a large nasal cavity that projects anteriorly, the presence of extreme bite force is evident, with a large brain and no true chin. All of this is reflective of a response to cold and chewing stress, and the use of the teeth as tools indicated by the forward placement of the jaws and the large size of the incisors. The Neandertal nose, located far from the brains, could have

functioned to warm cold air before it reached the lungs, and could possibly have served to

dissipate body heat during frequent periods of heightened activity. Post cranial features are also

indicative of adaptation to the cold in terms of the short and stalky body conformation and short

limbs with a barrel chest leading to a low surface area to body mass ratio thereby providing for

more effective insulation and thermoregulation, (Cavalli-Sforza *et al* 1986;Trinkaus 1986;

Wolpoff *et al* 1989;Frayer 1992;Rak *et al* 1994; Vandermeersch *et al* 1995). Diet may have been

a factor as well.

It is made clear by fossil evidence that early populations of the genus *Homo* incorporated

large amounts of animal material into their diet, (Wolpoff 1980;Bunn 1981;Leonard 1994). Some

have linked metabolic rates to brain size, and it has been argued that this not only correlates to,

but also is what initiated, other important anatomical and behavioral adaptations, (Bunn

1981;Milton 1987;Potts 1988; Leonard 1994). Population size would also affect overall physical

changes.

It was perhaps out of necessity that some interbreeding did occur among various populations

of intermediates, as small populations migrating out were susceptible to the rise of deleterious

mutations at greater rates due to bottlenecking. It is likely that some periods of varying degrees

of isolation occurred either due to environmental or cultural barriers giving rise to an increase in

inbreeding. This may have lead to some populations suffering from a high percentage of

deleterious mutations resulting in a population decrease. Breeding with neighboring

intermediates, when contact was available, would have been advantageous. Culture, meaning

tool use and subsistence systems, also mediated terms between individual populations and the

environment. Since behavior is generally considered genetically oriented, and passed on in a

way such that is considered heritable, then this too altered the evolutionary process by altering certain behavioral allele frequencies within the population, thereby increasing or decreasing overall fitness, (Bunn 1981, Potts 1988;Leonard 1994; Felsenstein 1999; Howard 1999,2000).

Gene expression is known to be subject to the actions of hormones, the production which is affected by a variety of external as well as internal factors. Variations, which arose in the actions of various androgens, impacted on hominid evolution perhaps to a greater degree than current data would indicate. The first cold of the Pleistocene could have resulted in greater selection for an increase in hormones affecting bone density, overall body mass, and brain development. This may have resulted in the more robust morphologies seen in many Neandertal populations. Subsequent Pleistocene climatic changes and differential reproduction changes affecting the dehydroepiandrosterone (DHEA, a testosterone precursor) and testosterone ratio may have also played a role in the extinction of a variety of such populations. Changes in these hormonal substances may produce allometric and behavioral changes that have been identified in modern populations (Nadler *et al* 1980;Winter *et al* 1980; Davis 1996; Howard 1999,2000).

The return of the cold in the later Pleistocene could have returned selection for increased DHEA. Neandertal habitat was characterized by relative containment. The cold and containment increased DHEA and testosterone in Neandertals. These hominids also continued to increase in brain size. The teeth and facial structures and the brain development of Neandertal are exaggerated because of the increase in testosterone and DHEA. The large teeth and brains are indicative of the fact that there was sufficient DHEA to affect a variety of tissues. This large brain, however, was increased in posterior regions, not in anterior regions, a development consistent with the onset of early puberty. High blood levels of DHEA and testosterone could lead to an early puberty. Increased androgen receptors in the brains of individuals with higher testosterone also increases use of DHEA for brain growth during childhood, thus accelerating the onset of puberty. The brain structures that control puberty mature early. This shortens the time for hypothalamic stimulation of testosterone production by the gonads. As testosterone-target-tissues grow and begin to increase in their use of DHEA, intra tissue competition for available DHEA increases. Therefore, early puberty reduces the amount of available DHEA for growth of anterior parts of the cerebrum. Large bodies, (increased muscle mass and bone density), and early puberty combine to reduce anterior brain development. The diaphyses of the radius, cervical vertebrate and the femur have all been shown to correlate dimensionally to higher or lower levels of DHEA. High levels of DHEA have been shown to increase bone size and prognathisism. Continued cycling of cold in the upper Pleistocene and changes in containment areas for Neandertals selected for hominids with ratios of DHEA and testosterone so different from other hominid populations, as this was necessary to produce the physiological changes needed for cold adaptation. High levels of testosterone have also been associated with a decrease in immune capabilities, an effect, which is exacerbated when trauma is involved, since the testosterone levels interfere with the release of corticosterone (Frayer 1994;Wiehmann *et al* 1996 Offner *et al* 1999;Ishikama *et al* 1999; Howard 1999; 2000). This may have played a role in the extinction of Neandertal populations, but also perhaps in the extinction of a variety of neighboring populations. This may have led to the eventual emergence of populations becoming adapted to warming climates and the periodic reduction of the percentage of individuals with higher testosterone possible in later populations. While isolation limited population interaction, or perhaps prevented some, a degree of variation existed and persisted. Selection pressures would have acted on these, thereby causing the development of local populations. Evolution, along with some degree of gene flow, could have promoted the spread of morphological and behavioral changes across a single species, evidence of which is present in populations existing today.

The Neandetals most likely interbred here and there with neighboring populations of other intermediate forms as migrations continued, and as geography permitted. Did the apparent

disappearance of these populations result from a cultural and or biological absorption? It is also

possible that there was not only gene exchange, but cultural exchange in terms of fire and tool

use as well. However, a genetic divergence seems to have possibly occurred about 500-

600,000ya, most likely due to isolation and inbreeding and/or this idea of absorption, perhaps

only of individuals. Neandertals shared similarities with other intermediate populations in terms

of basic anatomy, social organization, and perhaps similar subsistence strategies in terms of

subsistence patterns, thereby allowing for the classification of Neandertals as a subspecies of

Homo sapiens. Another possible explanation for the extinction of Neandertal populations would

be that the same increased competition that may have led to populations migrating out of Africa

may also have led to ongoing and rapid adaptations in populations inhabiting the same ranges as

other populations of intermediates, leading to a state of maladpativeness and an overall decrease

in overall population fitness for the adaptively lagging population, (Webster 1987; Wolpoff *et al*

1989; Dykes 1995;Vandermeersch *et al* 1995;Goodson *et al* 1996;Balliard 1997; Holliday 1997).

When talking about the evolutionary life history of the genus *Homo*, it is important to identify

that material (fossil, molecular and cultural) which is important to include within the genus.

There are many who question the taxonomic placement, the inclusion of Neandertals within this

genus, while others are advocates for it. Various theoretical models concerning the evolution of

anatomically modern humans have been developed as a result of this.

Theoretical Models for the Evolution of the genus *Homo*

The oldest of the seemingly anatomically modern *Homo sapiens* fossils are found in Africa (

Jebel Irhoud), and are dated to about,150-250,000ya, (Brauer *et al* 1989;Stringer *et al* 1993).

They most likely evolved from one localized population of origin for the species. They demonstrate clear evidence of having evolved from local African populations of the time, and most researchers agree that this local evolution had to occur in at least one place. The argument about the origins of modern groups of humans though centers on what happened next, because the next step either involved: local change as separate groups in each part of the world, or that the so-called moderns evolved in Africa and migrated out replacing existing, less -evolved populations. These two theories dominate the argument. However, there have been a total of four variations taking form. These models will be presented in overview with the emphasis being placed on the two dominant ones along with the supporting fossil and genetic evidence that supports them.

The African Replacement Model or (RAO) argues that modern humans first arose in Africa about 10,000ya and spread from that continent throughout the world. This model has been rigorously defended by Christopher Stringer, and subsequently supported by genetic (mtDNA) evidence presented by Cann and Stoneking (Stoneking and Cann 1989;Cann1992; Cann *et al* 1994). The African Hybridization and Replacement Model is basically the same RAO, although it allows for a greater or lesser extent of hybridization between the migrating population and the indigenous premodern populations of Europe (Aiello 1993). This position has been supported by Gunter Brauer. Next is the Assimilation Model. It differs from the previous models in denying replacement, or population migration, as a major factor in the appearance of modern humans (Aiello 1993). This model emphasizes the importance of gene flow, admixture, and changing selection pressures resulting in directional morphological change. Thus, local speciation is possible in the rise of moderns. This is most notably supported by Smith and Trinkaus, both of whom continue to question cladistic methodology, (Aiello 1993). Finally, The MultiRegional Evolution Model contrasts the African origin model for moderns. Instead, it emphasizes the role of both genetic continuity overtime and gene flow between contemporary populations in arguing that moderns arose not only in Africa, but also in Europe and Asia as well during the Middle Pleistocene, (Aiello 1993). Milford Wolpoff (*et al,* 1988, 1989) has been an adamant supporter of this particular model.

The theoretical problems begin with defining exactly which genera are valid, and then deciding who belongs where based on the parameters for defining the genera in the first place. The problem is that when material is discovered and described, it is sometimes overzealously given a new taxonomic designation, often based on extremely fragmentary material. These generic names sometimes go unresolved and fuel the debates for long periods of time. As a result, loss in the essence of human evolution is inevitable, as long lists of genera are then compiled and presented in various forms. Solutions to taxonomic questions based on current theoretical and cladistic models therefore are difficult and ambiguous, (Aiello 1993, Rak 1993, Park 1998).

The patterns of regional genetic variation predicted by the two main models are in themselves distinct. The MRE predicts that the similar, yet locally distinct in terms of phenotypic expression of traits, evolutionary transformations occurred across Africa, Europe and Asia, with local populations of intermediates gradually becoming more "modern" at a rate varying with locale. Local differences in drift, selection, and access to gene flow accounted for the variations found between populations. Some restriction in gene flow allowed for, or accounted for, differentiation of gene pools, leaving some populations to exhibit a greater degree of phenotypic variation than others. The RAO, on the other hand predicts different patterns in variation in African populations compared with those found elsewhere. Variation should be greatest within African populations, based on their supposed earlier divergence, and assuming predominantly neutral genetic change. They should be quite distinctive in terms of allele frequencies from the non-

African populations and transitional fossils would not have been found to occur outside the African area of origin. This would justify the population replacement argument to some extent. Transitional, or blending, shown in the fossil record in Europe, Asia, and Africa dispute this model, as the human species demonstrates a great deal of morphological variation, with genetic variation being minimal, and there has been little protein variation found between human populations. As much as 84 percent of protein polymorphism in human populations results from variation among individuals within a population, so variations between populations is often small in comparison to variations within them, (Trinkaus 1982, 1986; Cavalli-Sforza *et al* 1986; Cann *et al* 1987;Stringer *et al* 1988; Wolpoff 1985,1986, *et al* 1989, Aiello 1993; Rak 1993;Park 1999). Genetic data has been used to explain many questions about the history of *H. sapiens*, as it has been used to look at the relatedness between living as well as non-living human specimens.

The technique of G banding on chromosomes, a process of labeling them so that any chromosome can be recognized by the pattern of stain has allowed for comparisons to be made when working with genetic samples between and among populations. Chromosomes change with time due to various types of mutations, which, unlike genetic mutations, can be detected by this method at such a level. Conclusions about relatedness can be inferred from this, but problems do exist. Karyotypes of very closely related species are difficult to separate out, and these conclusions also rely upon the idea that change is equal to time. What isn't known is the rate of chromosomal change and whether or not it is constant. Various methodologies for genetic analysis have emerged to address such problems,(Caan *et al* 1987; Jarnick *et al* 1998;Stoneking *et al* 1998). Sequencing of amino acids provides direct information on which amino acids are changing between species. Myoglobins and hemoglobins are used specifically for such studies, as they can be quite similar between related species, and therefore act as indicators of evolutionary similarities. Again rates of change for amino acid substitutions caused by mutations can be a problem, and loss of conserved information (conserved regions) can occur as well. Conserved regions, or regions of high variability, are valuable landmarks in gaining both structural and functional roles of DNA regions (it makes a difference what genes are being observed at the time). The conserved regions dictate the identity of a structure, as they must contain a certain amino acid structure. The variations tell a different story, as mutations can survive in areas where they will either not cause impairment to function or enhance function. Then the degree of divergence between one species and another can be inferred to some degree, therefore the story of biological function is written in the degree to which the sequences of a particular functional molecule (such as myoglobin) are similar. Since identification of conserved regions is critical to such analyses, it is important to note how easily they can be lost in the extraction or handling of aDNA from fossils, (Cann and Stoneking 1987;Brauer 1992;Jarnick *et al* 1998;Stoneking *et al* 1998).

DNA has conserved regions, which do code for such things as calmodium (regulator of calcium levels in cells) and hemoglobin, along with many segments, which do not code for anything, and are most likely not conserved. DNA hybridization is a technique, which boils DNA to split it into strands, lets it cool and then it reassociates into a double helix. If it is mixed with the DNA of another species hybrids will form if at least 80 percent or so of the sequences match. In employing radioiodine, a measure of the degree of hybridization can be made. The melting curve of this will indicate the degree of hybridization and therefore the degree of similarity between the two species. By melting the DNA (separating the double strand) and then measuring the rate or renaturation, it is possible to estimate the complexity of an organism's genome. Altering temperature, pH, or ionic strength of the DNA will cause denaturing (or melting it from a double strand to single strands). The strands of two species are brought together in an assay analysis which measures the renaturation (hybridization) rate through the UV resorption process. The combination of the assay curve plotted over time in relation to the temperature changes is the melting curve. Low changes in temperature during hybridization indicate the hybrid DNA to be very similar to ordinary DNA and therefore indicative of the degree of relatedness between species. Overall mutation rates as well as estimated divergence dates between species are determined through such a comparison. The dates provided using these mutation rates seem to bring the genetic data more in line with fossil evidence, (Can *et al* 1987; Bauer 1992; Jarnick *et al* 1998; Stone king *et al* 1998). The one problem with this is that the process requires a good amount of DNA, about 500 nucleotides at once, and making use of random bits doesn't work, so restricting enzymes are used which are cut out of parts of functional DNA for the analysis. This then is really the use again of protein sequences, and another problem is the fact that there is such high intrapopulation variability in comparison to interpopulation variability, that the use of nuclear DNA is rendered ineffective as a diagnostic tool. Mitochonrial DNA was turned to as an alternative. Present and heritable through the female, it made tracing maternal lineages a possibility. The idea was to collect samples from populations as widespread as possible and fairly isolated in order that the present would bear some resemblance to the past, (Cann and Stoneking 1987). It was this work which led to the sequencing of mtDNA from 147 individuals from seven major subcontinental areas (Sub-Saharan Africa, Asia, Australia, Papua and New Guinea, Europe, North Africa, and the Middle East). They found 133 different types, only seven of which represented more than one individual, and of these groups, Africans were found to have the highest degree of variability. They then constructed a phylogenetic tree of all types. Starting from their one ancestor, there is an early split into a lineage which is exclusively African, but doesn't contain all Africans, and another which gradually splits to include all other Africans, and then progressively the other five groups, (Cann and Stoneking 1987). Basically this material offers support for the RAO model, as it states that there was an ancestral African origin, more than one migration out of Africa, and based upon mutation rate calculations the first ancestor lived at about 140-290,000ya with the first migration occurring at about 23,000-180,000ya. This is perhaps overstated and overly accepted. This "Eve" DNA type from a single population of perhaps several hundreds of thousands of females with most having left perhaps no descendents that lived long enough to breed, and given the small size of most populations and the fact that males hardly ever transmit mtDNA, then it is probable that these five types could actually be reexamined and reduced to a single type. This supports the single species hypothesis and makes sense of the little genetic diversity seen between all human populations,(Cann and Stoneking 1987; Lum 1989; Vigilant *et al* 1991; Jin *et al* 1995;Burrows *et al* 1997; Stoneking 1993; Harding 1997;Parsons 1997;Wolpoff *et al*, 1989; Jorde *et al* 1998).

The work done in 1997 by Krings *et al*, utilized extracted mtDNA from small pieces of the humerus of a Neandertal discovered near Dusseldorf in 1856. Its DNA sequence was more similar to that of moderns than either was to chimp mtDNA, indicating probable genetic contact between varieties of intermediate populations, and demonstrating support for the MRE model. Still, the mtDNA sequencing is out of the established normal parameter for human variability by about an 8 base-pair difference in the particular piece of mtDNA studied. This supports the ROA model, or does it? The problem with the use of some of this DNA is that when only segments, and smaller ones at that, are being utilized the problem exists of merely being a matter of protein sequencing, and not accounting for the probable missing pieces of functional DNA. What is being used for sequencing must be accounted for. Protein sequencing presented the problem of again knowing rates of mutational changes, (Brauer *et al* 1992; Kriggs *et al* 1997;Zietkiewicz *et al* 1998;Quintana-Murci *et al* 1999). In terms of the two main models (RAO and MRE) a closer look at both genetic as well as fossil evidence is in order. Evidence for the

RAO model comes from two sources, the fossils and the tools left behind some 100-300,000ya.

Fossil evidence shows no indication that hominids older than approximately 2mya existed at any other location outside of Africa , so it is fairly certain to say that the point of origin of the species was indeed Africa and spread from there. Early populations of the genus *Homo* in Africa spread to Asia by 1.6mya and to Europe by 0.8mya. They were classified as *erectus* material, yet most likely were only a sampling of the intermediates of the original population, as were later Neandertals. The *erectus* fossils found in Africa have been determined to be similar to those of specimens of the same species found in Java and China. The African remains have been found to be less robust, as would be expected as a result of variability and environmental adaptations by local populations of intermediates. It is uncertain just how much gene flow occurred between the African, Asian and European populations, yet these were all groups of people living at times when climates were in flux. In all populations of a species there are traits which differed between its members and which are heritable. Due to environmental changes over time, these traits change due to selection pressures leaving the population with a different genetic composition and phenotypic appearance. There can be a change in allele frequencies which show up in the physical evidence of fossils of the same species. Microevolution has very little to do with speciation, but is much more important in the need for a population developing local adaptations. Macroevolution on the other hand involves several mechanisms. Take a single population, and if its given geographical environment separates this population into two or more others, they will each diverge genetically if the local environments call for differing adaptations. After the populations have been separated for a period of time their genes may become so differentiated such that if reunited breeding may no longer be possible. The above mentioned fossil record however would contradict the idea that Neandertal total isolation occurred at all or for a long enough period of time to cause them to speciate from *H. sapiens*. Another mechanism would be the difference between two sets of nuclear genes, the structural genes that encode the adult's enzymes and structural proteins such as muscles; and the regulatory genes that operate only during early development to effect the general body plans of the embryo.

One regulatory gene affects the action of many structural genes by determining whether the structural genes make their proteins or not. It is possible to see that a single mutation in a regulatory gene can have far more important consequences for an organism's structure than a single mutation in a structural gene, because one regulatory gene mutation can completely alter the types of proteins produced and the timing of production of each protein. In addition to being able to make large structural alterations, mutations in regulatory genes can also produce changes rapidly, over say only a few generations, (Bauer 1992;Smith 1994; Jurmain *et al* 1995; Krings *et al* 1997; Linfield *et al* 1999; Quintana-Murci *et al* 1999). Thus regulatory genes are now recognized as a major mechanism for macroevolutionary change with little doubt that they have operated along with geographical separation to generate the rapid and extensive changes in human lineage over the past four million years. It is also important to keep in mind the previous mention of how these regulatory genes can also be subject to hormonal influences.

Though there is not a complete fossil record for the important period of 500-200,000 ya, the fossil evidence suggests that the lineages leading to moderns as well as Neandertals have been separate for a longer time than would be allowed by the MRE model. Dates of origin from the fossils for Neandertals is more than 200,000ya and for moderns about 100,000ya, and the morphological characters of these groups has held relatively constant throughout that period. Though it is believed from the fossil evidence that they each did in fact evolve from the original population in Africa, this technically lends support to a single species hypothesis with the resulting variety due to local adaptations. The fossil record also suggests that these two groups were together in two locations for perhaps several thousand years, in Israel 100,000ya and in

France 35,000ya. It is not certain whether the two were actually in the same exact location at the same time, but it is possible because of the nature of the Chatelperronian industry, which may have been a Neandertal tool assemblage with Cro Magnon influence. It is not known, for certain, if interbreeding occurred, but this suggests at least some cultural contact between groups. Gene flow would certainly support and explain the idea of a parallel evolutionary path, or continuity among intermediate populations evolving in different locations at different rates. Both DNA from the nucleus and the mitochondria, as previously noted, have been used to either refute or support the fossil record.

Genetic analysis indicates that Africa is clearly the source of all hominid DNA and that other groups split off from this root at some point, or points, in the past. The DNA evidence shows some important features: a single origin, intrapopulation variation higher that interpopulation variation suggesting that a greater gene pool was available to migratory populations, and finally, that the split from the original population occurred about 120,000ya with perhaps another European /Asian split at about 60,000ya. Mitochondrial DNA puts a different spin on this.

Over a long period of time random mutations crop up on mtDNA alleles, leading to the current variation seen among populations, basically the farther back in time, the greater the variety seen. Critics of Cann *et al* (1987) have maintained that the variability in mutation rates in mtDNA was underestimated, and that the model of RAO implied in some way that the *erectus* specimens suddenly became extinct after only one of its populations evolved into moderns in Africa. It doesn't account for the possible interbreeding between any of the intermediate populations which is suggested by the fossil data. The problem of only utilizing a small segment of the DNA has also been pointed out by critics of Cann for the reasons previously stated, that mutation rates are not known, and that the use of such segments (restriction) results in the use of the protein sequencing. Considering both the fossil and genetic data, it would seem that a more critical analysis of the genetic information can be made than would be the case for the fossil evidence.

In summation of the two models (RAO and MRE), the Neandertals in Europe and the Near East were most likely migratory populations of the original African one. Moderns, or so-called moderns, are evident as far back in Africa as 250-300,000ya, suggesting that differences did occur and at various times and places based on local cultural and biological needs. Neandertals perhaps suffered from a degree of isolation leading to far too many deleterious mutations to maintain a fit population thereby causing immune deficiencies and possibly a maladaptive state. Also, competition may have led to the isolation. Previously it had been noted that a degree of sharing of tool technology could be found among Neandertals and Cro Magnon populations. So did Neandertals become absorbed or die out? Neandertal populations probably had many individuals die off; entire local populations may have become extinct, while many others could have become absorbed by other intermediates again accounting for evidence of blending in the fossil record. Genetic data does suggest that all humans today are closely related, and that genes have all been derived from single source rather than parallel sources. Discoveries in Portugal in November of 1998 of a juvenile Neandertal-like skeleton and modern -like characteristics is just one of the fossil examples that exists to demonstrate gene flow between intermediate populations within the genus *Homo,* therefore supporting the classification of Neandertals as a subspecies of *Homo sapiens*, (Trinkaus *et al* 1986, 1989; Wolpoff *et al* 1986,1989;Cann *et al*l 1987;Wood *et al* 1991; Brauer 1992; Stein *et al* 1993; Stringer *et al* 1993;Smith 1994; Jurmain *et al* 1995; Treisman 1995;Krings *et al* 1997; Linfield *et al* 1999; Park 1999;Quintana-Murci *et al* 1999).

Growth and development as well as gene regulation in evolutionary time has become a focus of the appearance of anatomical patterns in hominid evolution. Morphological differentiation and climatic adaptation could be attributed to the function of the previously mentioned regulatory genes, providing a more in-depth explanation for the variation seen anatomically, as well as culturally, among various Neandertal and other intermediate hominid populations during the Pleistocene, while, at the same time, lending support to the MRE model.

Dehydroepiandrosterone and Hominid Evolution

The human hormone dehydroepiandrosterone (DHEA), a testosterone precursor, may enhance the effects of the hormone estrogen as well producing a more robust body confirmation and increased bone density in females and in males. It is produced in the adrenal gland, and though the most abundant of the steroid hormones. Its physiological implications are unclear. It has been suggested that DHEA can have dramatic effects on bone growth, as well as impacting negatively on the immune system. DHEA has also been shown to increase in an individual in response to cold environments. In hominids, the effects of testosterone on brain size, teeth, and other tissues are due to changes in the body's production of DHEA, and it has been suggested that DHEA is involved in the growth and maintenance of all tissues, particularly the brain, and is the leading cause in the formation of the more robust body confirmation seen in Neandertal populations, (Perel *et al* 1981; Tagliaferro *et al* 1986;Verdonck *et al* 2000; Howard 1999,2000).

Dehydroepiandrosterone and its sulfate ester are neuroactive and both are imported into the brain from the circulatory system, and are readily absorbed by the nervous system. Cold weather selects for a greater increase in both production and absorption of DHEA in order to bring about the necessary physical adaptive changes. The enlargement of neural tissue during primate evolution, and subsequently hominid evolution, may have been due to increased absorption and production of DHEA. Since DHEA is a precursor to both androstenedione and testosterone, growth prior to greater testosterone production is reliant upon DHEA levels. It is believed that DHEA leading to increased testosterone was advantageous due to its ability to optimize transcription and replication of DNA, (Perel *et al* 1981; Tagliaferro *et al* 1986;Baulieu *et al*

1999; Wickings *et al* 1999;Verdonck *et al* 2000; Howard 1999,2000). Therefore it may be that testosterone's advantage is that is directs DHEA use for genes that are targets (including bone and muscle tissue) of testosterone action, as it increases DHEA use. When testosterone increased use of DHEA for the brain, the available DHEA for growth and development of canine teeth was reduced, leading to the belief that the brain may be the primary tissue affecting hominid evolution due to its ability to capture DHEA to a greater extent than other tissues. When the brain increases in size, the canines tend to decrease, as brain tissue simply takes more DHEA up for use therefore the canines do not grow as large in response, (Perel et al 1981; Tagliaferro *et al* 1986;Baulieu *et al* 1999; Wickings *et al* 1999).

The cold periods of the Pleistocene directly caused changes in hominid evolution, which resulted in the development of the robust body conformations seen most notably in Neandertal populations. These cold periods may have selected for individuals that produced more DHEA. Rats injected with DHEA have shown increases in body weight, develop stronger body composition, show a more efficient utilization of dietary energy through the subsequent impairment of fat synthesis along with the promotion of fat-free tissue deposition and resting heat production (Tagliaferro *et al* 1986). This effect of DHEA is due to increased thermogenesis (Bobyleva *et al* 1993). Individuals who produce more DHEA derive more heat from the same nutrition. As cold decreased availablenutrition, those individuals who could derive more benefit from sparse nutrition had a survival advantage.

During the first cold period of the upper Pleistocene this more robust body type has been documented, to some extent, in some australopithecine material as well, explaining the noticeable variation in hominids existing prior to the Neandertal populations migrating out and adapting to various other environments. It also offers some explanation for the existence of what were perceived as the non-European Neandertals. Teeth and facial structures are also seen as representatives of testosterone-target-tissues. The available DHEA, not utilized in thermogenesis, caused an increase in the size of both teeth and facial structures. As the production of DHEA and testosterone increased during this time period, increases in growth continued in the brain and began to affect growth of the body overall. Unfortunately, the effects of these hormones increased, and continued even as the climate grew warm. There are no selection pressures known to reduce these levels in response to increased environmental warming. Reduced use of DHEA for themogenesis, however, increased the availability for body and brain growth. This applies to both males and females, and it is possible that what has been perceived as a decrease in sexual dimorphism in some specimens of the genus *Homo* could have actually been the signs of an increase in testosterone in both sexes. Although, there may be other interpretations with respect to the australopithecines and the existence of preexisting variation needed for the selection process overall. Bone mineral density has been shown to increase in response to increases in testosterone levels, and a direct correlation in bone formation and increased levels has been noted,(Nadler *et al* 1980;Ozasa *et al* 1982;Perry *et al* 1996; Sarrel *et al* 1998;Offner *et al* 1999; Howard 1999,2000).

The return of the cold later in the Pleistocene may have brought about selection for increased DHEA. Neandertal populations in Europe experienced some degree of isolation, based upon climatic (geographical) and/or cultural differences, perhaps experienced the development of increased levels of testosterone due to increased DHEA, again effectively increasing brain size. The teeth, facial structures, and brain development of Neandertals are exaggerated due to increased levels of these hormones indicating ample amounts were available to share with other target tissues. Since the brain was seen only to increase in posterior regions and not anterior regions, an early puberty occurred in these populations, again as a result of increased hormonal levels. Increased androgen receptors in the brains of individuals with higher levels of testosterone would increase use of DHEA for brain growth during childhood. This accelerates the onset of puberty because the brains structures controlling puberty mature early. This shortens the time before hypothalamic stimulation of testosterone production to the gonads. As testosterone-target-tissues grow and begin to increase their use of DHEA, competition for available DHEA increases. Therefore, early puberty reduces available DHEA for growth of anterior parts of the cerebrum. Large bodies and early puberty reduce final anterior brain development. That is, early puberty reduces the time of basic growth and development of the brain that occurs under the influence of DHEA. Assuming then that sufficient DHEA is available, too much testosterone increases bone size and prognathism and too little has the opposite effect, (Ozasa et al 1982; Perry et al 1996; Sarrel et al 1998;Offner et al 1999; Ishikama et al 1999;Verdonck et al 1999;Howard 1999,2000).   Continued cycling of the cold during the upper Pleistocene and changes in containment areas selected for hominids with different ratios of DHEA and testosterone which would coincide with these changes. Reduced nutritional availability or quality will slow the pace of puberty. Increased nutritional supplies and quality should favor those with early puberty. The percentage of individuals with high testosterone levels would decrease, and average size of the forebrain would increase, this most likely began with the emergence of the genus Homo, (Offner et al 1999; Ishikama et al 1999;Verdonck et al 1999;Howard 1999,2000).   High levels of testosterone have also been shown to compromise the immune system. The effects are especially dangerous when trauma is involved, and males have been shown to be associated with a dramatically increased risk of major infections following trauma, (Offner et al 1999). The maintenance of immune function by androgen deficiency does not seem to be related to changes in the release of corticosterone. The negative effects of testosterone on immunity could increase the probability of infectious epidemics that could radically change the percentage of individuals with high testosterone levels in a population. In terms of the Neandertals, this could have meant that not only were they more susceptible to infections perhaps introduced by neighboring populations, or by the altering environmental effects on microbial life, but also more at risk for infection due to traumas suffered for whatever reason. Neandertals would then loose many of the better-adapted individuals in colder settings thereby effectively decreasing overall fitness. During warmer periods, this increase may also have acted as a selective disadvantage, and further was possibly a mechanism in decreasing overall fitness leading to the extinction of certain select populations. Only if enough individuals with higher levels remaining would then be able to aid in the recovery of the select allele frequencies over time, therefore not only were climatic cycles possible, but perhaps testosterone cycles as well. This would offer an explanation for the eventual decline of body size in both males and females by the end of the upper Paleolithic lasting through the Neolithic. This might explain much of the variation in anatomical features found in hominids of the Pleistocene, and again support the taxonomic placement of Neandertals as a subspecies of *Homo sapiens*, (Frayer 1984; Wiehmann *et al* 1996;Bachrach *et al* 1999;Baulieu 1999; Howard 1999,2000).

Final Note on Neandertals

Neandertals were first discovered in 1856, and clarification has still not been established concerning the relationship between their populations and the genus *Homo*. Some feel they were a subspecies demonstrating local adaptations to cold environments, while other still classify them as separate species which became either hybridized with moderns and/or was forced into extinction due to the overpowering by more intelligent moderns. Even genetic evidence has yet to show any conclusive results, which might shed light on this ongoing debate. By the early 1900's a large number of Neandertal remains had been uncovered, primarily in France. Rather than assisting in finding a resolution, the debate heated up. Anatomical morphologies and tool archeological remains only add to the classification and origin conflict. To this day some classify Neandertals as a subspecies within *Homo sapiens*, while others list it as a separate species. Genetically, some feel that there are remnants of Neandertal DNA still found in modern DNA, while others dispute this saying that there is too vast a difference in genetic sequencing. Milford Wolpoff is perhaps the leading advocate for what is known as the Multiregional Evolutionary Model. This model explains Nenadertals as one of many intermediate populations of *Homo* that migrated form Africa breaking away from the main population of a single species in search of food, better climatic conditions, and perhaps even less resource and habitat competition. The variability is explained in terms of merely a case of regional adaptation with distinct alleles for particular traits were selected for in order to assist in drawing the population into a state of optimum fitness. This is also given as the explanation for all intermediate populations, culture and biology evolved based on the needs of the given environments of each population. After all, some have found that anatomically modern humans existed as early as 300-250,000ya, and many have found similar phenotypic traits in Neandertals found in Europe in non-European specimens, meaning variability within the species made it possible for selection pressures to act on preexisting alleles which were advantageous later in colder environments. Another factor accounting for similarities would be that some believe that interbreeding took place between neighboring populations of intermediates, and that the various minor genetic differences are a result of some populations or even perhaps individuals, becoming isolated either from their own population or from neighboring populations at various points in time based upon climatic and environmental events.

Other than Wolpoff, Ofer Bar-Yosef also has felt that interbreeding occurred between Neandertals and other intermediates at the time the advances of the glaciers took place forcing

Neandertals to move further south into areas inhabited by other intermediate populations. This too would also offer an explanation for the findings for the non-Europeans again as well. The similarities found showing a blending between traits of various intermediates, including Neandertals. Retreating glaciers allowed then for *Homo sapiens* to also follow Neandertals back into northern regions. Still there are others, citing anatomical changes in the more recent Neandertals, think they evolved independently into early Europeans. Wolpoff suggests that the more appropriate approach would be to investigate the possibility that Neandertals are ancestors of some Europeans. There is a good possibility that Neandertals evolved form a population of early *Homo sapiens* in Africa that may have experienced some periods of population or individual isolation. Evidence shows Neandertals displayed body types and cultures that adapted them to the cold. Recent evidence has pointed to evolution as a consequence of differential gene regulation, in continuous and relatively comparable genomes, resulting mainly form chronological differences in production of the androgenic hormones, testosterone and dehydroepiandrosterone (DHEA). The cold periods of the Pleistocene selected for individuals with more DHEA, which interacted with subsequently higher levels of testosterone that varied according to behavioral advantages. The two principle events of hominid evolution would then be increased testosterone in populations affecting body size, bone density, tooth and facial feature development, brain size and regional development, as well as having a negative effect on immune system responses by causing an increased risk for the development of infections. Neandertal fossil finds were considered to be in areas of prime climatic advantages during the Pleistocene in the middle of Europe, therefore supporting the idea that other neighboring populations could have been nearby and interbred and exchanged cultural ideas. Smith, Trinkaus and Wolpoff argue for the possibility the populations interbred, that they originated in Africa, and over time there were various migrations out by small populations establishing a subspecies based on having to adapt to local geographic conditions, as did all intermediate populations of the species of origin.

Many fossil finds, of juveniles in particular, (Stringer *et al* 1985,;Tattersall 1986;Trinkaus 1986, *et al* 1989),demonstrate anatomical as well as some genetic similarities between Neandertals and other specimens of *Homo*, and their mtDNA is not completely outside of the modern human range. It would appear that the differences between Neandertals and other *Homo* specimens may not be as significant as it is believed to be by those attempting to still find support for the replacement model. It would appear that as more fossil and genetic evidence surfaces, support is gained for the regional continuity model and the idea that Neandertals indeed ought to be classified as a subspecies of the genus *Homo*. Increasing support for this model also supports the idea that extinction was perhaps a result of maladaptation over time due to increased competition with neighboring populations leading to an eventual increase in deleterious mutations and eventual maladaptation to their environments on both a cultural and biological level.

Increasing evidence continues to support the idea that there was an original population of one species with small groups that broke off at various times adapting to local geographies and experiencing genetic and cultural contact with neighboring populations. As this evidence grows, it may very well cause a sweeping reassessment of current theories on the history of Neandertals.

Comparative Anatomy and Osteology of Extant Primates

Overview of Primate Anatomy

The primates are a diverse eutherian (placental) group with an extensive life history throughout geological time. Transitional primate-like mammals, such as tree shrews and Purgatorius, first appeared by the end of the Mesozoic Era (65million years ago), (Cartmill, 1975;Ciochon *et al*, 1987). The proliferation of mammalian forms, along with the emergence of these primate-like mammals, from this point on coincided with the extinction of many other life forms, particularly the dinosaurs, (Cartmill, 1975). There are approximately 193 living primate species placed into 13 families. The smallest of the primates living in the wild is the pygmy lemur, which weighs about 30 grams, the largest is the gorilla, weighing up to 175 kilograms, (Napier *et al*, 1967;Kavanaugh, 1983).

Primates, or their tree shrew-like predecessors, (Cartmill, 1975), were an opportunistic niche filling mammal who radiated once the populations of dinosaurs virtually disappeared. The depopulation of the planet likely opened up many new niches, which accounts for the rapid increase in mammalian species following the Cretaceous-Tertiary boundary. Primates , or their tree shrew-like progenitors were one of these opportunistic, niche-filling mammals. An early theory of primate evolution, proposed by F.W. Jones, (1916), relates primate characteristics of grasping hands and feet, orbital frontation (increased binocularity of vision) and enhanced cognitive processing capacity to the challenges of arboreal life,(the arboreal theory of primate evolution). Primates radiated in arboreal habitats, and many of the characteristics by which they are recognized today (shortened rostrum, and forwardly directed orbits associated with stereoscopic vision, relatively large braincases in comparison to other mammalian groups, opposable thumbs, unfused and mobile radius, and ulna in the forelimb and tibia and fibula in the hindlimb) arose as adaptations for life in the trees or are primitive traits (from those of the last known common ancestor). Such traits were retained for the same reason, they served as adaptations to the new lineages. Several species began to leave the trees early on as environmental changes occurred and adapted to a terrestrial life, and yet still retained many of these features. Some primate evolutionary trends have included a shift towards stereoscopic vision, shortening of the muzzle, a decreased dependency on olfactory senses, nails rather than claws, sensitive pads for gripping, and an increase in brain size. Primates are recognized based on sets of characteristics of the skull, teeth and limbs, making identification and distinction of remains (fossil or otherwise) possible,(LeGros Clark, 1961;Napier *et al*, 1967, Cartmill, 1975;Kavanaugh, 1983, Fleagle, 1988).

Dentition

Primates have four types of teeth, incisors, canines, premolars, and molars. Incisors are like shovels that cut and lift food, canines are pointed and not only cut, but can also act as weapons in most primates. Premolars and molars have large crown surfaces that shear and grind food during mastication, (Herzkovitz, 1977; Ankel-Simons, 1999). The number of teeth and types are summarized by listing only the elements on one side from incisor, canine, premolar to molar (mesial to distal) for each jaw. Therefore, the primitive mammalian formula of 3.1.4.3/3.1.4.3 represents a probable mammalian ancestor with a total of 44 teeth. However, no living primate has more than three incisors or three premolars on each side in either the maxilla or mandible. Prosimians and Platyrrhines have three premolars, while Catarrhines have only two. Third molars are often absent in some primate genera. Once a tooth is lost (due to the adaptation process) in a species it is usually not reproduced again, meaning that descendants tend to possess fewer teeth than their ancestors did, (Savage *et al*, 1986;Fleagle, 1988;Ankel-Simons, 1999). Primate dentitions vary widely both between genera and between species within a genus. In most primates the canine teeth are much longer than the other teeth. In humans the size of the canines are reduced and the ends are blunt. Short canines allow for increased side to side movement of the jaw. The short canines in humans are a functional consequence of the enlarged cranium and associated reduction of the size of the jaws. The canines also became reduced in size in response to the increase in vegetation in the diet. In primates, canines function as both defense weapons and visual threat devices. Interestingly, the primates with the larger canines with respect to body size (gorillas and gelada baboons) both have basically vegetarian diet, demonstrating that the larger canines are primarily a defensive tool, (Savage *et al*, 1986; Kay *et al*, 1994).

The teeth of primates vary considerably. The dental formula for the order is 0-2/1-2, 0-1/0-1, 2-4/2-4, 2-3/2-3 = 18-36. The incisors are especially variable. In some forms, most incisors have been lost, although all retain at least 1 lower incisor. In others, the incisors are intermediate in size and appear to function as pincers or nippers, as they commonly do in other groups of mammals. In some, including most strepsirhines (see next paragraph), the lower incisors form a toothcomb used in grooming and perhaps foraging. In the aye-aye (Daubentoniidae), the incisors are reduced to 1 in each jaw and are rodent-like in form and function. Canines are usually (but not always) present; they vary in size, including within species between males and females. Premolars are usually bicuspid (bilophodont), but sometimes canine-like or molar-like. Molars have 3-5 cusps, commonly 4. A hypocone (4th cusp) was added early in primate history, and the paraconid (part of the triangular cusp pattern of early mammal-like reptiles) was lost, leaving both upper and lower teeth with a basically quadrate pattern. Primitively, primate molars were brachydont and tuberculosectorial , but they have become bunodont and quadrate in a number of modern forms. Living primates are divided into two great groups, the Strepsirhini and the Haplorhini. Strepsirhines have naked noses, (wet rhinarium) with separated nostrils, lower incisors forming a toothcomb for grooming and no plate separating orbit from temporal fossa. The Haplorhini (higher primates) do not posses a toothcomb, and do have a plate which separates the orbit from the temporal fossa, (Savage *et al*, 1986; Kay *et al*, 1994; Fleagle *et al*, 1995).

The tooth itself consists of a crown, the portion covered with enamel, and a root of dentine covered with cementum. The interior of the tooth is the pulp chamber that consists of soft cellular tissue. The primary mineral in both bone and dental structures is crystals of apatite (a form of calcium phosphate). Dentine is a bone-like substance (about 75% mineral) but enamel is much more heavily mineralized (96% mineral by weight), (Ankel-Simons, 1999). Cementum, whose composition differs only slightly from that of dentine, attaches the tooth to the periodontal ligament and provides an interface between tooth and surrounding bone, while enamel forms a crystalline cap over the actual working surface of the crown, (Ankel-Simons, 1999).

Primates have two sets of teeth during their lives, a deciduous set that is replaces by permanent teeth during childhood and adolescence. The Old World higher primate dental formulas are:

Deciduous:	incisors	canines	Premolar/molar	
Maxilla	2	1	2	
Mandible	2	1	2	
Permanent	incisors	canines	Premolar	Molar
Maxilla	2	1	2	3
Mandible	2	1	2	3

Note that in humans there are no deciduous premolars. Substantial individual variation occurs in tooth number, but most often the variant is the loss of a tooth at the boundary between tooth kind (incisor, canine, premolar, molar). Humans are also missing fourth premolars, leaving them with three or four. The greatest difference between humans and most other primates is in the canine teeth. Small , incisoral human canines do not project from the tooth row. Diastemas, gaps in the tooth row of the maxilla allow projecting mandibular canines to pass the opposing canine and incisor during occlusion. The maxillary canine passes the buccal side of its opposing third premolar, allowing the lingual surface of the canine to make contact with a blade-like sectorial surface of the premolar. Humans lack this large diastema and the human third premolar is non-sectorial. Human anterior teeth (canines and incisors) are greatly reduced in size compared to their early hominoid (ape-like) ancestors, and the incisors are positioned close to a transverse plane that passes through the canine teeth. Other primates, the chimpanzee in particular, have incisors which are positioned well forward of this plane. As a result, the parabolic shape of the hominoid dental area contrasts with the more sharply defined U shaped area of the others, such as the chimp. Also, human molars tend to be more rounded due to a softer diet, along with an increase overall in foodstuffs and foodstuff variety ,which requires more of a grinding action as opposed to cutting or shearing. Human molars are also more compact in comparison to other primates, and the occlusal molar surfaces of human teeth are relatively flat and quickly become even flatter over time, (Napier *et al*, 1967;Kavanaugh, 1983; Savage *et al*, 1986;Fleagle, 1988).

Skeletal Elements

The skull consists of 28 bones that are described as the bones of the calvarium (supporting and surrounding the brain) and those of the face. Tiny bones of the middle ear, the conchae and vomer in the nose, and the ethmoid of the orbital cavity are not easily seen. Although usually not considered part of the face, the hyoid bone form a skeletal element for the larynx. In addition to the openings that represent the eyes, nose, mouth, and ears, the skull has numerous foramina for the passage of nerves and vessels. The bones of the cranium (the skull minus the mandible) are joined by irregular sutures that fuse after growth ceases, (Hartman, *et al*, 1933; Kavanaugh, 1983;Ankel-Simons, 1999).

There are several major differences between human skulls and those of other primates. Within those of the non-human primates however, there are even more degrees of variety to be found. The brain volume of the human is about three times greater in terms of the endocranial volume than that of non-human primates. This is to accommodate the larger brain in humans, which ranges in anatomically modern humans from roughly 1200cc -1400cc. The size of the face relative to the braincase provides and indirect feel for relative brain size. In humans, there is a negative correlation between face size and the neurocranium. The face is relatively small in comparison to the neurocranium, meaning that humans have a large brain even when taking into consideration their body size, though they do not possess the largest brains overall. The largest brain is possessed by the mammal, the sperm whale (Physeter catadon), which is about seven times the size of a human's, (Norwak, 1991). Though the human's is the largest among primates. Distinguishing differences between apes and monkeys can only be accomplished when precise data is available pertaining to brain volume and body size. However, apes have the relatively larger between the two, (Ciochon *et al*, 1987; Aiello *et al*, 1995;Ankel-Simons 1999).

The larger human brain volume also means that the calvarium is large relative to the size of the face. Though not yet well understood, it is believed that as the brain expands during the early phases of embryonic development leading to vesicle formation, it appears to flex on an axis around the pituitary fossa. This produces flexion of the cranial base, a downward shift in the posterior part of the cranium, and a forward rotation of the foramen magnum. (Swindler *et al*, 1973; Hill, 1974; Ankel-Simons, 1983; Marieb, 1997). The larger range of the human calvarium allows for sufficient surface area for the attachment of the larger and stronger temporalis. In mammals, this muscle evolved to become advantageous in allowing the jaws to capture and subdue prey, (Swindler *et al*, 1973; Hill, 1974; Ankel-Simons, 1983).

There is an expansion of the occipital lobes, and the cerebellum swells. This leads to the correlation with the formation of an asymmetrical cranial venous sinus system. In the human, a superior sagittal sinus drains venous blood in a transverse sigmoid route to the internal jugular veins. Enlarged occipitomarginal sinus systems typical of humans are infrequent in other primates. The human middle cranial fossa expands with the enlargement of the temporal lobes of the brain expanding the midsection of the calvarium outward above the petrous portion of the temporal bone and the glenoid fossa. In others, such as anthropoids and hominoids, the calvarium tends to be narrower than the cranial base, but the anatomically modern human calvarium has its maximum width high on the parietal eminences rather than low toward the cranial base. The anterior cranial fossa expands with the human frontal lobes. Therefore, anatomically modern humans lack the postorbital constriction that occurs in primates such as the chimpanzees, and modern humans also exhibit frontal eminence (forehead) above a much less obvious supraorbital torus (eyebrow ridge). Olfactory nerve tracts pass through the cribriform plate, a structure in the middle of the ethmoid bone. In humans , the median plane of this plate is the site of a process, the crista galli, which is absent in some other primates, (Hartman *et al*, 1933; Swindler *et al*, 1973; Kavanaugh, 1983; Ankel-Simons, 1983,1999).

The ethmoid bone itself developed in response to the overall evolution of the vertebrate skull. The ethmoid bone is exceedingly light and spongy, and cubical in shape. It is situated at the anterior part of the base of the cranium, between the two orbits, at the roof of the nose, and contributes to each of these cavities. It consists of four parts: a horizontal or cribriform plate (previously mentioned) forming part of the base of the cranium; a perpendicular plate, constituting part of the nasal septum; and two lateral masses or labyrinths. During development, the cribiform plate is received into the ethmoidal notch of the frontal bone and roofs in the nasal cavities. Projecting upward from the middle line of this plate is a thick, smooth, triangular process, the crista galli, so called from its resemblance to a cock's comb. The long thin posterior border of the crista galli serves for the attachment of the falx cerebri. Its anterior border, short and thick, articulates with the frontal bone, and presents two small projecting alæ, which are received into corresponding depressions in the frontal bone and complete the foramen cecum. Its sides are smooth, and sometimes bulging from the presence of a small air sinus in the interior. On either side of the crista galli, the cribriform plate is narrow and deeply grooved; it supports the olfactory bulb and is perforated by foramina for the passage of the olfactory nerves. The foramina in the middle of the groove are small and transmit the nerves to the roof of the nasal cavity; those at the medial and lateral parts of the groove are larger, and the former transmit the nerves to the upper part of the nasal septum, the latter ones transmit the nerves to the superior nasal concha. At the anterior part of the cribriform plate, on either side of the crista galli, is a small fissure which is occupied by a process of dura mater. Lateral to this fissure is a notch or foramen which transmits the nasociliary nerve; from this notch a groove extends backward to the anterior ethmoidal foramen. The perpendicular plate is a thin, flattened lamina, polygonal in form, which descends from the under surface of the cribriform plate, and assists in forming the septum of the nose. It is generally deflected a little to one or other side. The anterior border articulates with the spine of the frontal bone and the crest of the nasal bones. The posterior border articulates by its upper half with the sphenoidal crest, by its lower with the vomer. The inferior border is thicker than the posterior, and serves for the attachment of the septal cartilage of the nose. The surfaces of the plate are smooth, except above, where numerous grooves and canals are seen; these lead from the medial foramina on the cribriform plate and lodge filaments of the olfactory nerves. The Labyrinth or Lateral Mass (labyrinthus ethmoidalis) consists of a number of thin-walled cellular cavities, the ethmoidal cells, arranged in three groups, anterior, middle, and posterior, and interposed between two vertical plates of bone. The lateral plate forms part of the orbit, the medial, part of the corresponding nasal cavity. In the disarticulated bone many of these cells are opened into, but when the bones are articulated, they are closed in at every part, except where they open into the nasal cavity. The upper surface of the labyrinth presents a number of half-broken cells, the walls of which are completed, in the articulated skull, by the edges of the ethmoidal notch of the frontal bone. Crossing this surface are two grooves, converted into canals by articulation with the frontal; they are the anterior and posterior ethmoidal canals, and open on the inner wall of the orbit. The posterior surface presents large irregular cellular cavities, which are closed in by articulation with the sphenoidal concha and orbital process of the palatine. The lateral surface is formed of a thin, smooth, oblong plate, the lamina papyracea (os planum), which covers in the middle and posterior ethmoidal cells and forms a large part of the medial wall of the orbit; it articulates above with the orbital plate of the frontal bone, below with the maxilla and orbital process of the palatine, in front with the lacrimal, and behind with the sphenoid. In front of the lamina papyracea are some broken air cells which are overlapped and completed by the lacrimal bone and the frontal process of the maxilla. A curved lamina, the uncinate process, projects downward and backward from this part of the labyrinth, where it forms a small part of the medial wall of the maxillary sinus, and articulates with the ethmoidal process of the inferior nasal concha. The medial surface of the labyrinth forms part of the lateral wall of the corresponding nasal cavity. It consists of a thin lamella, which descends from the under surface of the cribriform plate, and ends below in a free, convoluted margin, the middle nasal concha. It is rough, and marked above by numerous grooves, directed nearly vertically downward from the cribriform plate where they house branches of the olfactory nerves. These nerves are then distributed to the mucous membrane covering the superior nasal concha. The back part of the surface is subdivided by a narrow oblique fissure, the superior meatus of the nose, bounded above by a thin, curved plate, the superior nasal concha; the posterior ethmoidal cells open into this meatus. Below, and in front of the superior meatus, is the convex surface of the middle nasal concha; it extends along the whole length of the medial surface of the labyrinth, and its lower margin is free and thick. The lateral surface of the middle concha is concave, and assists in forming the middle meatus of the nose. The middle ethmoidal cells open into the central part of this meatus, and a sinuous passage, termed the infundibulum, extends upward and forward through the labyrinth and communicates with the anterior ethmoidal cells, and in about 50 per cent of all skulls (Warwick *et al*, 1973;Picq *et al*, 1989 , 1994) it is continued upward as the frontonasal duct into the frontal sinus. The ethmoid is ossified in the cartilage of the nasal capsule by three centers: one for the perpendicular plate, and one for each labyrinth. The labyrinths are first developed, ossific granules making their appearance in the region of the lamina papyracea between the fourth and fifth months of fetal life, and extending into the conchæ. At birth, the bone consists of the two labyrinths, which are small and ill-developed. During the first year after birth, the perpendicular plate and crista galli begin to ossify from a single center, and are joined to the labyrinths about the beginning of the second year. The cribriform plate is ossified partly from the perpendicular plate and partly from the labyrinths. The development of the ethmoidal cells begins during fetal life. The ethmoid articulates with fifteen bones: four of the cranium : the frontal, the sphenoid, and the two sphenoidal conchæ; and eleven of the face: the two nasals, two maxillæ, two lacrimals, two palatines, two inferior nasal conchæ, and the vomer. In addition to the usual centers of ossification of the cranium, others may occur in the course of the sutures, giving rise to irregular, isolated bones, termed sutural or Wormian bones. They occur most frequently in the course of the lambdoidal suture, but are occasionally seen at the fontanelles, especially the posterior. One, the pterion ossicle, sometimes exists between the sphenoidal angle of the parietal and the great wing of the sphenoid. They have a tendency to be more or less symmetrical on the two sides of the skull, and vary much in size. Their number is generally limited to two or three; but more than a hundred have been found in the skull of an adult hydrocephalic subject, (LeGros Clark,1961; Warwick *et al*, 1973;Picq *et al*, 1989, 1994; White, 1991;Ankel-Simons, 1983,1999; Lieberman, 2000; Marieb, 2000).

The swelling out of the calvaria in the temporal and occipital area associated with the expansion of brain volume produces an apparent forward shift in the relative position of the foramen magnum. The human foramen magnum lies on a line, the bitympanic line that connects the most inferior points on the lateral end of the right and left tympanic plates. Most often with other primates, such as the chimpanzees, pygmy chimps and bonobos, the foramen magnum lies behind this line, (Hartman, *et al*, 1933; Swindler *et al*, 1973). Muscles that attach to the cranial base serve to position, move, and stabilize the head and cervical vertebrae. From a functional perspective, the skull is also the most superior point of origin for muscles (such as the M. trapezius) that provide movement and stability to the back and shoulder. As a result, some of the powerful muscles that attach to the skull have large attachement surfaces. In non -human primates the smaller skulls , combined with more robusticity and muscle tend to produce a stronger sagittal crest for attachement. The human occipital is a cup-like bone with a noticeable eminence, the external occipital protuberance. It is marked by nuchal lines and markings that represent attachment surfaces. The nuchal crests of other primates tend to be more extreme due to their quadrupedal posture. The purpose of nuchal crests is for attachment of neck muscles,. In non-human primates, especially with respect to the gorilla, these crests need to be larger and thicker than those of other primates due to a greater need to hold the head upright in the maintenance of a quadrupedal type of habitual locomotion pattern. In humans, such crests along the occipital region are present, but are slightly different in that they do not need to hold up the cranium as completely as they do in other primates. In humans such crests are smaller and act more as origin sites for less robust muscles while still aiding in stabilizing the head and cervical vertebrae. This is due to an upright, bipedal type of locomotion and posture which reduces the stress and robusticity of this region. There is also a good amount of sexual dimorphism found in the occipital region. In males an external occipital protuberance is present, while it is generally absent in females. The occipital bone itself also presents much rougher in texture in males than in females. Due to the flexion of the cranial base with expanded endocranial volume , the superior nuchal line is lower on the human skull. Also, the human mastoid process, the insertion surface of the sternocleidomastoid muscles which originate at the sternum/clavicle , is more distinct and separated from the outline of the occipital by a digastric fossa. These processes (one on each side) are inferiorly-projecting nubs of bone, placed laterally on the basicranium, just posterior to the coronal plane, which serve as attachment points for the major muscles of lateral head rotation (looking from side to side) and abduction (tilting head forward after looking upwards). (Hartman, *et al*, 1933; Swindler *et al* , 1973; Kavanaugh, 1983;Ankel-Simons,1983, 1999;Picq *et al*, 1989, 1994;White, 1991).

The smaller human incisivum, a homologue of the chimpanzee premaxilla, joins with the maxilla near the time of birth. In most primates, the premaxilla is epanded to support the large and forward projecting incisors. It joins with the maxilla postnatally and the suture between it and the maxilla become fused. The projection of the face beyond the calvaria is greater in most other primates (prognathic) than in humans (orthognathic). The craniofacial angle, the angle between the most anterior point on the maxilla, the most anterior point of the sphenoid bone and the most anterior point of the foramen magnum, is used to determine the extent to which the face projects beyond the neurocranuim, because this angle can not easily be measured on intact skulls, (Hill, 1974). Due to this problem, the angle that a sellion-prosthion line makes with a Frankfort plane (an imaginary line from the lower margin of the eye socket to the notch above the tragus of the ear) serves as a way of making measurement of facial projection easier. This plane passes through porion, right and left orbitales. The porion is the most lateral and superior point of the external auditory meatus. The orbitale is the lowest point of the infraorbital margin. The sellion is the deepest point in the hollow beneath the glabella in the median plane. The glabella is the most anterior point in the median plane on the supraorbital torus. The prosthion is the most anterior point on the maxilla in the median plane,(Swindler *et al*, 1973;Hill, 1974;Ankel-Simons, 1983). The human mandible is reinforced by what is known as a bar of bone, the mental protuberance, that strengthens the symphysis (joining of the right and left halves of the bone). The lingual , or posterior, surface of the symphysis bears a pair of tubercles that represent muscle attachment sites for the M. genioglossus and M. geniohyoid. The ape mandible lacks a mental protuberance and is reinforced by an inferior transverse torus (or simian shelf). Viewed from above, the contrast is shape of the dental arcade becomes distinct, (Swindler *et al*, 1973;Hill, 1974;Ankel-Simons, 1983; White 1991).

The human hyoid is a U-shaped bone just above the larynx. The stylohyoid ligament attaches the lesser horn of the hyoid to the styloid process of the temporal bone. Generally, these styloid processes point to the lesser horns of the hyoid bone. In the chimpanzee, the hyoid is expanded anteriorly to make room for a larygeal air sack, and is located higher in the neck, (Hill, 1974;Ankel-Simons, 1983,1999). The human hyoid is a U-shaped bone just above the larynx.

The stylohyoid ligament attaches the lesser horn of the hyoid to the styloid process of the temporal bone. The thyrohyoid membrane is not part of the larynx but connects the cartilaginous skeleton of the larynx to the hyoid bone. This membrane connects the body and greater wing of the hyoid bone to the upper border of the lamina and superior horn of the thyroid cartilage. The posterior free border of this membrane is thickened to form lateral thyrohyoid ligaments and the midline part of this membrane is thickened to form the median thyrohyoid ligament. The internal laryngeal nerve and superior laryngeal vessels perforate this membrane. The thyrohyoid membrane forms the lateral boundary of the piriform recess. Generally, these styloid processes point to the lesser horns of the hyoid bone. In the chimpanzee, the hyoid is expanded anteriorly to make room for a larygeal air sack, and is located higher in the neck, (Hill, 1974;Ankel-Simons, 1983,1999; Marieb, 2000).

In primates, most possess a sagittal crest to anchor the temporalis, as it acts to pull the lower jaw towards the skull. In humans, the crest is absent due to both dietary changes, (cooked foods, softer diets of both plants and animals) and an increase in brain size, so the temporalis (which is located within the temporal fossa) is attached to the top process (coronoid) of the mandible to the side of the braincase just below the zygomatic process. Unlike the masseter, whose attachment allows for a broad mastication application, the temporalis' focused attachment (form) allows for a focused function, that is to close the jaw until the teeth come into contact. The temporalis' evolutionary role has been to aid in grabbing prey with the canine teeth, which is why the temporalis' strength is greatest when the mandible is open about 20 mm. Once the teeth come together, the temporalis has built in resistance (the maxilla), and can contract at many times its necessary intensity. In carnivores (those animals with prominent canine teeth) and both human and non-human primates then, the temporalis is the strongest and most efficient muscle of mastication, as some of the fibers of the lateral part of the temporalis muscle originate from the temporal fascia. The temporal fascia attaches to the zygomatic arch and braces the arch against the pull of the masseter. (Swindler *et al*, 1973; Ankel-Simons, 1983; Aeillo *et al*, 1995; Marieb, 1997).

The Jaw

The four primary muscles of mastication are the temporalis, masseter, (both synergists, as both are involved in closing the jaw), medial pterygoid, and lateral pterygoid. The masseter and the pterygoids form a u-shaped sling for support,and act synergistically in order to provide power for the crushing and shearing . Other smaller muscles in the neck, including the digastric, attach to the hyoid and assist in stabilizing or lowering the jaw. The temporalis arises again , from the side of the braincase and from the overlying temporal fascia and inserts on the coronoid process of the mandible. On the cranium the muscle leaves scars, the temporal lines, indicating the extent of its attachment. The temporal fascia, which is anchored to the zygomatic arch, is suggestive of the membrane that closed over the fenestra in the synapsid skull. The convergence of fibers from the two origins gives the muscle a bipennate arrangement that is appropriate for a muscle with great strength and limited distance of contraction. The synapsid condition is characterized by the presence of a single temporal fenestra bordered minimally by the jugal, postorbital, and squamosal. The quadratojugal and the parietal occasionally contribute to the edge of this fenestra. By comparison with diapsids, this fenestra can be called a lower temporal fenestra. All early members of Synapsida, the taxon that includes mammals and their extinct relatives, had a synapsid skull, but the temporal fenestra has been drastically modified in mammals by ventral processes of the frontal and the parietal that occlude the temporal fenestra. The location of the old fenestra is still visible between the zygomatic arch, the orbit, and the dorsal part of the skull, but it is no longer a hole in the skull, (Osborn, 1903; Heaton, 1979; Basmajian , 1985, *et al*, 1985; Hylander, 1985).

The primary action of temporalis is elevation of the mandible in the act of biting or chewing. From a lateral view, the fibers from the muscle appear to be fan-shaped and are different lengths, indicating that they have multiple functions. The posterior fibers are important in retraction of the mandible. It has been suggested that the more horizontal posterior fibers, which are aligned with the anterior dentition, are more important in incisive biting, while the more vertical anterior fibers allow a greater biting force for the molars in the back of the mouth, (Basmajian , 1985, *et al*, 1985; Hylander, 1985; Aiello, 1990).The mass and strength of the temporalis, in general, is dependent on the surface area from which it can arise on the side of the cranium. In most primates, including the great apes and australopithecines, the strong development of the muscle in adult males leads to an expansion of the origin as far as the sagittal suture. Additional surface area on the bone is provided by a sagittal crest rising vertically on the skull and a nuchal crest posteriorly on the occipital bone. Such crests are absent from the genus Homo in part because of the reduced size of the musculature and in part because the extreme expansion of the brain and braincase provide abundant surface area for muscular attachment. The massive masseter, arising from the zygomatic arch is functionally simpler. Attaching to the mandible below and posterior to the molars, the masseter, assisted by medial pterygoid applies its forces primarily in molar chewing, (Basmajian, 1985, *et al*, 1985; Hylander, 1985).

During the course of hominid evolution, one of the most striking series of changes involved the development and loss of a powerful chewing complex . This pattern contradicted the long held notion that evolution was linear, from an apelike ancestor to anatomically modern humans, and that any so - called "missing link" would be something of an average between chimps and humans. Instead, it is now known that the evolutionary pathway can wander, following adaptative strategies unique to a particular epoch or species. Primitively for mammals the mandible is composed of right and left dentary bones joined at the midline symphysis by ligaments. Such a joint can provide some movement, if desired, but it can also weaken the jaw under heavy chewing stresses. Thus in anthropoids (and many other mammalian groups) the symphysis is fully fused and immobile. This appears to relate to a larger body size and tougher diet, (Du Brul et al, 1954; Biggerstaff, 1977; Basmajjan, et al, 1985; Demes et al, 1988). Modern apes and monkeys typically reinforce the symphysis with ridges of bone on the inner curvature of the chin called a mandibular torus or simian shelf. This reinforcement behind the joint resists the actions of the pterygoid muscles in squeezing the condyles together (Daegling 1993). Various primates have also strengthened that area by making the jaw deeper near the midline (Hylander, 1985 ; Ravosa 1991).Modern humans have replaced the shelf with an external protuberance, the chin. The prominent human chin distinguishes the species from other living primates and from all earlier hominids. It has been argued that the development of a chin is the only character that consistently distinguishes anatomically modern humans from nearly all archaic populations of hominids, (Neandertals perhaps an exception), (Pope 1992). Several explanations have been put forth regarding the evolution of the primate chin, not all of them conflicting (DuBrul et al, 1954). As the jaw shortened, it potentially "crowded" the tongue and floor of the mouth. The lower part of the mandible was therefore reduced less than the tooth-bearing region, leaving a jutting chin. Developmentally, these two parts of the mandible appear to be under independent controls (Biggerstaff , 1977). Developmental models may tell how the chin developed, but not why. Mechanical models of chewing stresses can answer that (DuBrul et al, 1954; Scott 1963). Human chewing produces vertical bending stress and also twisting and shearing stresses near the symphysis (Hylander, 1985). It is reasonable to conclude that the chin represents a reinforcement of the bone to resist fracturing, although it may not be possible to identify any one type of stress over others as responsible for its unique shape, (DuBrul *et al*, 1954; Scott 1963).

The human mandible is reinforced by what is known as a bar of bone, the mental protuberance, that strengthens the symphysis (joining of the right and left halves of the bone). The lingual , or posterior, surface of the symphysis bears a pair of tubercles that represent muscle attachment sites for the M. genioglossus and M. geniohyoid. The ape mandible lacks a mental protuberance and is reinforced by an inferior transverse torus or simian shelf.. This is due to the human chin. Human chins have a forward projection , the mental protuberance and mental tubercles (knobs), which often all extend farther forward than the lower teeth. Some living apes and early human ancestors have poorly developed mental knobs. Viewed from above, the contrast in the dental arcade shape becomes that much more distinct. The pongids have a rectangular shaped palate with very large canine teeth (left picture). In order to be able to close their mouth, there must be a gap (diastema) for the opposing tooth between the incisors and the canine. The fully modern human palate is parabolic in shape, rather than rectangular (right picture). The canines are relatively smaller so that the teeth can occlude without a gap for their placement. The lower jaw (mandible) must have the same general shape as the palate in order for the teeth to occlude. The rectangular shape of the pongid palate is represented in the mandible as well, including the diastema (left picture). Once again, the pongids have a rectangular shaped tooth row, whereas modern humans (right picture) have a parabolic mandible shape. The pongid mandible has a bony extension lingual to the incisors, called the ? simian shelf. ? Fully modern humans lack this bony extension, (Swindler *et al*, 1973;Hill, 1974;Ankel-Simons, 1983; Aiello *et al*, 1990; White , 1991).

Vertebral Column

The human vertebral column consists of 33 vertebrae divided into five functional regions, (Marieb, 1997). Seven cervical vertebrae are easily identified by their transverse foramina, and form the skeletal region of the neck. The articular surfaces between each are flexible in depending upon degrees of mobility in each region. The twelve thoracic vertebrae are only mobile within the coronal plane (frontal or horizontal plane) and function to support the ribs. Five more robust vertebrae are located in the lumbar region in the lower back and are tightly articulated in order to be able to withstand the weight of the torso. Five sacral vertebrae fuse to form the sacrum, the component of the axial skeleton in the pelvis. The sacrum is at the bottom of the spine and lies between the fifth lumbar segment and the coccyx (tailbone). The sacrum is a triangular-shaped bone and consists of five segments (S1 - S5) that are fused together and connect to the pelvis (ilium) and form joints called the sacroiliac joints. Finally, four caudal vertebrae extend downward from the sacrum. In adults these caudal vertebrae join to form the coccyx, a hidden human tail that partially blocks the inferior pelvic aperature, (Swindler et al, 1973; Marieb, 1997). If viewed from the side, humans have a series of four curvatures. The dorsal outline cervical and lumbar regions are concave while the thoracic and sacral regions are convex. The forward curve of the lumbar region is called lordosis and the backward like curvature of the thoracic region is called kyphosis. Most other non-human primates tend to have one more thoracic , one less lumbar, and one less caudal vertebrae than do humans due to being quadrupedal as opposed to habitually bipedal. The brachiators also would have the additional thoracic vertebrate as well due to the need for greater upper body support. Both the chimpanzee and humans normally have seven cervical vertebrae and normally the combined thoracic, lumbar, and sacral regions consists of 22 vertebrae. Chimpanzees though lack the extreme curves of the human column, and the angle between the lumbar and sacral region is more acute, (Hill, 1974;Marieb,1997;Ankel-Simons, 1983,1999).

Thorax

The human chest consists of 12 paired ribs that articulate with the vertebral column. Ribs I through VII attach ventrally to the sternum. Ribs VIII through X terminate in cartilage extensions that eventually fasten to the sternum. Ribs XI and XII have free distal ends. The sternum is composed of six flat bones that fuse in adults to three units, the manubrium body and xiphoid process. The individual bones that comprise the sternum are known as sternebra. These are connected to the ribs by the costal cartilage. A spear shaped manubrium is at the anterior end and the xiphoid process is at the posterior end of the sternum. The manubrim, also called the "handle", is located at the top of the sternum and moves slightly. It is connected to the first two ribs. The body, also called the "blade" or the "gladiolus", is located in the middle of the sternum and connects the third to seventh ribs directly and the eighth through tenth ribs indirectly. The xiphoid process, also called the "tip", is located on the bottom of the sternum. It is often cartilaginous (cartilage), but does become bony in later years. In some humans, especially older ones, the three units of the sternum may fuse to one another , (Swindler , 1973; Hill, 1974; Marieb 1997,2001). Many non-human primates have an added rib, or 13 pairs, this again being related to posture and locomotive patterns. The human rib cage is slightly broader for its depth, and the human thorax is less funnel-shaped. Many non-human primates have thoraxes more rounded when viewed cross sectionally since they are habitually quadrupedal and this provides for more effective muscle attachement, along with the differences in respiration due to posture, (Swindler , 1973; Hill, 1974; Ankel -Simons, 1983; Marieb 1997,2001).

Pectoral Girdle

The shoulder is formed by the humerus, clavicle and scapula. The clavicle attaches firmly to the manubrium and acts as a strut to hold the shoulder joint away from the chest. Distally, the clavicle articulates with the acromion process of the scapula, a large triangular flat bone at the back of the shoulder. The scapula actually forms the posterior part of the shoulder girdle, and its primary purpose is that of muscle attachment. The glenoid cavity of the scapula articulates with the head of the humerus. The most noticeable difference between humans and non-human primates is in the proportions of the scapula. Human arm strength, much less powerful in movements when in a raised position, is reflected in the shape of the scapula which provides attachment surfaces and lever arms for muscles. In the upright or erect posture, the scapula is somewhat enlarged and more vertical. The more enlarged/elongated the scapula, the greater the associated muscle strength. Maximized shoulder strength is a direct result of posture and type of locomotion, and therefore overall positioning of component muscles. The scapula acts to facilitate this positioning. In brachiators, the scapula spine is more oblique, larger and more dorsally placed, while in quadrupeds it is still somewhat larger due to posture and means of locomotion, but it is more laterally placed. (Swindler *et al*, 1973, Ankel-Simmons, 1983; Fleagle, 1985; Aiello,1990; Marieb, 1997).

Arms

The upper arm is a single bone, the humerus. While in the anatomical postion, (the most ubiquitous starting point from which to begin to describe anatomical features and positions), the arms are held rotated externally at the shoulders and the forearms fully supinated and therefore facing forwards. The body is not in the usual position adopted at rest. In terms of articulations

then, the forearm is formed medially by the ulna, which articulates firmly by a hinge joint to the

humerus, and laterally by the radius which is firmly attached to the hand. The radius pivots on

the humerus and ulna to supply pronation (hand rotation) and is attached to the ulna by powerful

interossesous muscles and ligaments. The head of the humerus (as well as the femur) is useful in

identification of gender of a mature unknown cadaver. If the maximum diameter of the head is

greater than 45mm, the individual is probably male. If the head is less than 42mm it is probably

female, (Preuschoft *et al*, 1993;Swindler *et al*,1973, Marieb, 1997).

Other non-human primates contrast with the human in this area as well. The human lacks the

robust lateral supracondyle ridge, a high and robust lateral epicondyle, and the steep, sharp,

lateral margin of the olecranon fossa. Brachiating and knuckle walking primates tend to have a

relatively longer forearm. They also have a radius and ulna which are more curved, and the distal

radius had a radiocarpal joint surface which diverges medially. The major differences between

human and non-human limbs are contrasts in relative proportion. The non-humans tend to have

larger , longer and more powereful arms, and shorter legs. The reasons for similarities between

the anatomical structure of the arms between both brachiators and knucklewalkers is due to the

fact that this portion of the body is used in each case for weight support, both differing from

humans since their arms are utilized for manipulation. (Preuschoft *et al*, 1973, Swindler et al,

1973, Ankel-Simons, 1983; Fleagle, 1985; Aiello,1990).

Arm or forelimb similarities between apes and humans reflect common origins. The humerus

has a wide trochlea in apes and humans compared to monkeys. The large trochlear ridge

contributes to stability between humerus and ulna, and frees the radius for pronation and

supination in a variety of elbow postures. The radial head is round in humans and apes, while in monkeys it is more elliptical. There are , however, a greater number of differences anatomically among the primates, particularly between apes and humans. In the distal humerus, the capitulum extends onto the posterior surface in apes, and gives the radius greater arc of movement. The lateral trochlea ridge is well developed and stabilizes the elbow joint in hyperextension when knuckle walking, and also prevents lateral dislocation. The olecranon fossa is found deeper in apes , with a steep margin and greater hyperextension. The lateral epicondyle and ridge is also more developed in apes and provides greater lever advantage of long extensors of wrist and fingers. The medial epicondyle and ridge are more developed in apes also, and serve to increase lever advantage of flexors of wrist and fingers, and provide an insertion area for extra pongid muscle (dorsoepitrochlearis - a tensor of arm fascia). The radius has a lateral curvature of shaft which is greater in apes, which allows for longer pronator muscles, as well as increased power of forearm and hand in pronation (in knuckle walking) in apes. Radial tuberosity is medial in apes and anterior in humans, and apes posses the greater mechanical advantage of biceps over other primates. The ulna's coronoid process and larger trochlea notch are more robust in apes due to the transmission of weight during knuckle walking locomotion, (Swindler et al, 1973, Andrews et al, 1976;Ankel-Simons, 1983; Fleagle, 1985; Aiello,1990).

Hands

The hand has three skeletal regions, the carpus, a series of eight carpals, form the wrist. The hand consists of five metacarpal bones. Phalanges form the skeleton of the fingers. The thumb, digit one has two phalanges (proximal and distal), while digits II through V have three phalanges (proximal, middle and distal). Most non-human primates differ in that their hands differ from the human's in terms of relative length of the digits. The thumbs are usually shorter, and phalanges tend to exhibit a more robust appearance, and are areas of insertion for flexor tendons, while metacarpals have transverse ridges across their heads that will limit dorsiflexion in order to reduce joint strain due to locomotor patterns, (Preuschoft et al, 1973 Andrews et al , 1976; Ankel-Simons, 1983;Marieb, 1997).

Hand morphologies and grips also exhibit a wide range of variation despite the fact that most species retain the primitive condition of five manual digits. These morphologies are related to postural and locomotor adaptations. However, feeding and grooming adaptations are important in the evolution of hand morphology. Both Strepsirhines (lemurs and lorises) and Haplorhines (tarsiers and anthropoids) show diversity in hand morphologies. The hands of anthropoid primates show less diversity than the hands of strepsirhine primates. Nevertheless, considerable diversity is evident in these forms. The main feature that distinguishes these forms is the size and shape of the thumb. Note that the human thumb is long compared to that of the chimpanzee . The human thumb is fully opposable (able to touch all the other digits tip-to-tip). This is not possible for chimpanzees, gorillas, and orangutans (although it is possible for some monkeys), (Biegert, 1963; Jungers, 1985; Aiello, 1990).

Pelvic Girdle

The pelvic girdle is formed by the sacrum, coccyx, and the two coxae. Each coxae is attached by ligaments to the sacrum and to each other at the pubic symphysis. A coxa is formed by the fusion of three bones, the ilium, the ischium, and pubis, which meet in the acetabulum (hip socket), (Marieb, 1997). The pelvic girdle supports the trunk on the legs and encloses and protects the pelvic viscera. Each half of the pelvic girdle is called the os coxae; where these two join in the front is the symphysis pubis. The pelvis is bowl-shaped, with greater and lesser pelves, separated by the pelvic brim. The os coxae have three distinctive features: iliac crest, acetabulum, and obturator foramen. The adult os coxae forms by the fusion of the ilium, ischium, and pubis. The ischial tuberosity is the knob upon which a body rests (sits). Other features include the greater sciatic notch, iliac fossa, and ischial spine. In humans the pelvis has a basin-like shape, while in apes it is longer and more narrow, also the females have larger birth canals. Therefore, there is a set of characteristics which can be used to identify both the gender and species of an unknown pelvis, (Andrews et al , 1976; Ankel-Simons, 1983;Aiello, 1990; Marieb, 1997,2001).

The pelvic inlet of the female is larger and has a greater absolute circumference. The superior ramus of the pubic bone is longer, increasing the pubic/ischium ration. The greater sciatic notch is wider and forms a longer angle. The increased pubic length and laterally displaced ischia result in a wider subpubic angle. The growth and remodeling of the pubis produces extra bone at the symphysis, leaving a concave inferior ramus, a ventral arc that represents a previous border of the symphysis, and a narrow inferior pubic ramus. The female pubic symphysis is likely to be longer in its superior-inferior diameter and smaller in its dorsal-ventral diameters than in the male. Females are more likely to have a well-developed preauricular sulcus, and those who have borne offspring may have pits along the dorsal border of the pubic symphysis. Since they have smaller femurs, females also have smaller acetabula, (Swindler et al, 1973; Roche et al,1996;Marieb, 1997). Such sexual dimorphisms in the pelvic region have an affect upon the carrying angle. The carrying angle is the angle at which the femur comes down and meets the knee. This does not occur in apes. In hominids a carrying angle is evidence for bipedalism. This angle is the amount of deviation from the perpendicular of the femur; 11 degrees in humans. The carrying angle provides better support and leverage for bipedal locomotion. Bipedalism is basically a problem of balancing the body while moving on one foot at a time. To maintain balance, the foot must be moved toward the midline of the body; the carrying angle allows this to happen. The angle itself also provides a clue as to gender, as the wider pelvis leads to an increased inward slant of the thigh and, therefore, an increased angle at the knee. Females also tend to have shorter limbs, especially the upper arms, and thus have less lever action with a resultant loss of power. Therefore, females also have an increased carrying angle at the elbow, (Andrews et al, 1976; Ankel-Simons, 1983;Aiello, 1990; Roche et al, 1996). The growth and remodeling of the pubis produces extra bone at the symphysis, leaving a concave inferior ramus, a ventral arc that represents a previous border of the symphysis, and a narrow inferior pubic ramus. The female pubic symphysis is likely to be longer in its superior-inferior diameter and smaller in its dorsal-ventral diameters than in the male. Females are more likely to have a well-developed preauricular sulcus, and those who have borne offspring may have pits along the dorsal border of the pubic symphysis. Since they have smaller femora, females also have smaller acetabula,(Swindler et al, 1973; Roche et al,1996;Marieb, 1997).

Differences in the pelvis reflect the differences between the habitual locomotion of humans and the quadrupedal movements of non-human primates. The pelvises of a non-human and human primate have radically different form and locomotor function. The human pelvis (bipedal) consists of two coxal bones plus the sacrum. Each of these elements has become stronger by the fusion of smaller elements. The sacrum represents the bodies of five different vertebrae whose movement has been eliminated. The coxal bone is produced by the union of three bones, the ilium, pubic, and ischium. The two coxal bones join on the midline in front at the pubic symphysis. The three parts of the coxal bone meet and fuse at the acetabulum, the deep socket for the hip joint. Above and below the acetabulum, the coxal bone extends as a flat surface for the attachment of broad muscles. The bone can be allowed to thin in the center because stresses occur primarily around the edges. The iliac blade above is so thin it can transmit light on a dry bone. Below the hip joint, the bone can be replaced by the obturator membrane, which serves the same function of muscular attachment. Along the superior edge, the iliac crest anchors muscles of the abdominal wall. Inferiorly, the ischio-pubic ramus anchors important muscles of the thigh. In anatomical position, the pelvis is oriented so that the two tips of the iliac crests and the pubis symphysis lie approximately in the same coronal plane. This directs the birth canal somewhat posteriorly and permits the pubic bones to give direct support to some of the viscera. The skeleton and musculature of the human hip and thigh show clear adaptations for bipedalism. A considerable proportion of the effort for balancing and shifting weight comes from the hip. Except for climbing behaviors, the most important hip movement occurs near full extension. In quadrupeds, full extension is impossible and gait movements center about a highly flexed position of the joint. Motion of the hip in abduction and adduction, although small in range, becomes far more crucial in a biped for balance than in a quadruped. The human pelvis is overall shorter, broader, and deeper than that of other primates and of mammals in general. Quadrupedal locomotion required hip flexors attaching more cranially on the pelvis and powerful extensors more caudally. An elongated pelvis provides a greater lever arm and thus more power for those muscles. In a human, cranial-caudal length is unimportant, while locomotion depends on ventral-dorsal attachments of the muscles. For this reason the ilium and ischium are both shortened. To gain some mechanical advantage from the shape of the coxal bone, however, it is not entirely aligned with the spine, but is tilted significantly in a sagittal plane. The short, tilted ilium also brings the sacroiliac and acetabular joints closer together to reduce shear stresses between these joints, (Haxton, 1947; Ankel-Simons, 1983;Campbell, 1985; Aiello et al, 1990).

As a result of brachiation, the pelvis has moved forwards at the top and back at the bottom. This is pelvic tilt. Both locomotor and postural behaviors determine the morphology of the pelvic girdle. The pelvis of hylobates is somewhat long or elongated and more narrow and broad, though not as broad as in pongids who only engage in modified brachiation. However, in both instances the ilia form flat blades, unlike the shorter , more broad and ventrally bent ilia of humans. The pelvis and sacrum artiuclation is narrower in brachiators, in non-human primates overall, so the space between the hip and the sacro-iliac articulation is also more narrow. The femur articulates at the proximal end of the hindlimb. The neck length and angle at which it articulates varies with each species' pelvic morphology and locomotor behaviors. The femoral head is more globular in most primates with the exception of the specialized leaping of prosimians, galagidae and tarsiidae. In these species the femoral head is tends to be more cylindrical, while in the terrestrial primates the femur is bent craniocaudally, (Stern et al, 1973;Jenkins, 1974;Ankel-Simons 1983). In quadrupeds the condyle articulation is more equal in terms of size, as with humans the two sides of the pelvic and femoral articulations tend to converge downward while they converge in an upward manner in pongids. As a result, the inner trochlear condyle of the lower femoral articulation is larger in humans and smaller in pongids, while the shaft of the femur is more narrowed and straight in lesser apes, yet more robust and forwardly tilted in pongids. In leapers such as tarsiers, the femoral head which articulates with the hip is more of a cylinder than of a globular shape, the long axis of the head is basically perpendicular to the long axis of the femur itself. The opposite is true of gibbons, since they are brachiators and movement is performed primarily with the arms, (Stern et al, 1973;Jenkins, 1974;Ankel-Simons 1983).

The relative width of the iliac blade is much larger in humans in comparison to non-human primates. As previously mentioned, in humans the ilium is broad and low while in other primates it is higher and more narrow. The humans have a slightly noticeable iliac pillar (thickening which extends from the iliac tubercle to the acetabulum). The humans also have an anterior inferior iliac spine. The human acetabulum is larger, again as a result of the larger femoral head, and the superior border of the acetabulum is reinforced so as to be able to sustain the stress of bipedal locomotion. The inferior border of the illium near the auricular surface forms the greater sciatic notch, where the articular surface is again larger in humans. The innominate bones of Papio again are elongated in a craniocaudal orientation, leaving the transverse diameter of the pelvis more narrow in quadrupeds. Their pubic symphysis includes both the pubis and ischium, and a sacrum composed of only two or three vertebrate. In brachiators, such as Pan, as in humans, the diameter is greater and there are five sacral vertebrate. Ischial tuberosities are more well developed and defined in cercopithecoids, this being linked directly to posture rather than locomotor behavior. In baboons there is a larger triangular development of the processes, which have rougher and more of a broad surface with an everted laterl margin. This forms the surface for the ischial callosities of Old World Monkeys. Such pads are utilized for sitting or sleeping in a seated position, (Haxton, 1947; Swindler et al, 1973; Ankel-Simons, 1983; Aiello et al, 1990) There has long since been an established relationship between anatomy and locomotor behaviors, but it is also such sitting postures as these which have contributed to the alteration of a species' morphology as well, and perhaps moreso in the case of the pelvic region since most primates engage in posture related behaviors more often than locomotor behaviors, (Morbeck et al, 1979).

Legs

The femur articulates ball and socket fashion in the acetabulum of the coax. The lower leg contains a large medial bone, the tibia, which articulates with the femoral condyles to form the knee. Lateral to the tibia, the fibula is a small , irregular, weight-bearing bone that provides attachment surfaces for muscles. Projections on the distal ends of the tibia and fibula, the medial and lateral malleolus, form a secure slotted proximal structure for the ankle joint. As the head in the humerus, a maximum diameter of 45mm or greater for the femoral head indicates a male gender. The human femur is longer than in most other primates, and usually has an elevated pilaster that supports the linea aspera down the shaft. The angles of the head, shaft, and condyles contrast greatly with those of the apes. Out of the three long bones of the hind leg, (femur, tibia and fibula) the femur is the most proximal and largest, and has three bony protrusions on the end: the caput femoris (head), which articulates with the hip bone., at the lateral side of the upper end is the trochanter major (greater trochanter) , where the gluteus muscles insert, and finally just below the femoral head is the trochanter minor (lesser trochanter). These trochanters are connected by a crest (intertrichanteric line). The backside also has a crest , the linea aspera, a line of intersection for several muscles. At the patella end there are two somewhat rounded inner and outer condyles. It is the axis between these which forms a right angle to the long axis of the femur, an axis which varies among primate species. The long axis never intersects the shaft in an ape femur. The femoral condyles of the human are larger and more elliptical than those of apes. The human femur has a large articular surface and mid-shaft circumference when compared to the arm than do apes. Apes have a similar ratio between femur length and tibia length but the ape leg is much shorter overall relative to the arm. The ape proximal tibia is smaller, less well supported by the shaft , and has condyles that arc more convex than is usual in humans, (Swindler et al, 1973; Kavanaugh, 1983; Roche et al,1996;Marieb, 1997, Ankel-Simons, 1983, 1999).

Among New World Monkeys the one feature that tends to have the most influence over the morphology of long bones of the hindlimbs is body size rather than locomotion, (Ankel-Simons, 1983). Therefore, two types of hindlimbs can be determined among these primates. The first is what Halaczek, (1972: in Ankel-Simons , 1983) calls the Aotus group, containing the Callitrichidae and cebids, all of small body size along with Pithecia. The second, or Ateles group, includes: Ateles, Brachyteles, Lagothrix and Alouatta . Cebus tends to demonstrate a morphological intermediate type. In the group of the smaller boy types, the skeleton of the hind leg is smaller, and shorter overall than the skeletal trunk. In the larger body type group, with the exception of Alouatta , the skeleton of the hind leg is longer than the skeletal trunk. In the smaller body type group again, the femur is also shorter than the tibia, and just the opposite in the larger body type group. The angle between the distal condyles of the femur is open laterally in the smaller monkeys and open medially in the large body type group, (Ankel-Simons, 1983).

Old World Monkeys exhibit a particular hindlimb morphology. Among Cercopithecoidea the length of the femur and tibia is shorter than the length of the skeletal trunk, Nasalis (the proboscis monkey) being the exception. The Femur and tibia are about a quarter longer in Hylobatidae, and among the pongids the hindlimbs of the orangutan are about equal to the skeletal trunk. Whereas leg length exceeds the trunk among both chimps and gorillas, this being attributed to their modified(knucklewalking) style of quadrupedal locomotion. In terms of robusticity, these long bones of the hind legs are about equal and somewhat robust in Cercopithecidae and Pongidae, while among the gibbons they are more slender, again due to differences in locomotor and posutural behaviors, (Swindler et al, 1973; Ankel-Simons, 1983; Kavanaugh et al, 1983; Roche et al , 1996).

Foot

Just like the hand, the foot has three anatomical regions. The seven bones of the tarsus form the ankle and proximal half of the foot. The middle portion of the foot consists of five metatarsals. Phalanges, the skeletal elements of the toes, have the same number and arrangements as in the fingers. The primary differences then between the human and non-human foot then are seen in the contrast between prehensile functions of the ape foot and the anatomy of bipedal striding in the human. The ape foot has an opposable hallux and long phalanges. The human foot has greatly reduced digits, with all metatarsals parallel and an increase in the lever arm of the tarsus for striding. In the human foot, a longitudinal arch provides a shock absorbing and weight distribution system. The orientation of the ankle joint allows the tibia to take a straighter path over the foot during walking, (Hill, 1972; Swindler et al, 1973; Kavanaugh, 1983;Roche et al,1996;Marieb, 1997, Ankel-Simons, 1983, 1999).

The tarsal bones are grouped as follows: A proximal row with calcaneus and talus, a distal row with meidl , intermediate, and lateral cuneiform bones, and the cuboid. Positioned between these two rows on the medial side is the navicular. The tarsal bones are larger and show greater variation in size. This is due to their direct correlation to the function of weight bearing, particularly in humans. Tarsal bone variation among primates shows that in both Papio and Pan the axis of the neck of the talus forms a 20 degree or greater angle with the articular surface of the tibia, whereas in humans it is far less pronounced, (Swindler et al , 1973). This difference is linked to the positioning of the hallux, abducted in brachiators and quadrupeds and adducted in humans. In addition, the articular surface of the tibia extends onto the dorsal surface of the neck in baboons, (Hill, 1970) but not in other primates. The calcaneus extends posteriorly into the tuberosity, and in humans it is positioned more inferiorly and posteriorly to form a heel. This posterior extension is more constricted in both Pan and Papio genera. The rest of the tarsal bones are morphologically similar with the exception of some minor size and proportional differences. In terms of proportions, there is an obvious shortening anteroposteriorly of the proximal and intermediate tarsal bones which LeGros Clark(1960) states is due to the forward displacement of the center of gravity in the great apes. The metatarsals are long and cylindrical with enlarged proximal and distal ends. In chimps the third metatarsal is the longest due to the functional axis of the foot, but in humans and baboons this axis is located between the first and second digits, making the second the longest, (Swindler et al, 1973). In terms of the hallux, it is only found in relation to the others in humans, where it is long, robust, and oftentimes the longest of the digits.

According to Schultz (1936) this length of the hallux is the result of the reduction of the other digits in the genus Homo, (Swindler et al , 1973; Ankel-Simons, 1983).

The talus articulates in the trochlear joint with the tibia and with the two malleolae. The malleolar facets are nearly vertical to restrain the talus within the joint. The calcaneus underlies the talus and supports body weight on the ground. The calcaneal tuberosity is the heel process and acts as a lever arm for plantar flexors of the ankle. This tuberosity developed an inferior process among hominoids and other climbing primates to enhance the action of short flexors of the toes. Humans have a unique second inferior process, more lateral, that increases the area of bone bearing body weight. One of the striking characteristics of the human foot skeleton compared with that of other mammals is the relatively large size of individual bones. The tarsals of a chimpanzee, for example, appear compact and strengthened with a thick layer of cortical bone. Human tarsals appear puffed up. This contrast can be understood as two alternative strategies for dealing with stresses of weightbearing. The thick cortical bone of the chimpanzee is strong and resists the pressures that might fracture it. The increased volume of human cancellous bone, otherwise filled with fat, is better able to absorb and dissipate the greater compressive stresses delivered during normal bipedal gait. The thin cortex permits some deformation that is padded by the fat-filled interior. This same pattern is also found in many other parts of the human skeleton, (Swindler et al, 1973; Kavanaugh, 1983;Roche et al,1996;Marieb, 1997, Ankel-Simons, 1983, 1999).

The longitudinal arch of the foot is a uniquely human construction. It plays a critical role in normal gait. The skeleton of the arch is the row of bones along the medial side of the foot: the calcaneus, talus, navicular, first cuneiform, and first metatarsal. Only the calcaneus and the head of the metatarsal normally bear weight, although the height of the middle of the arch is, and should be, variable. The longitudinal arch is a dynamic structure, increasing and decreasing in its height and strength during gait. Although the foot should be pliable and supple during passive weight-bearing and weight-acceptance phases, the arch must be rigid and strong in push-off. The ability of the foot to alternate between these two functional states is due to the conversion mechanism built into the major joints, (Cahill, 1965; Bruckner, 1992).

There are many articulations of the tarsal and metatarsal bones, but only a few have significant movements. Most are irregular gliding joints, permitted enough wiggle by their ligaments to relieve stresses and absorb shock, but not possessing identifiable axes of movement. Two important exceptions are the subtalar joint and the midtarsal joint. On the other hand, the tarsometatarsal joint at the base of the hallux is notable for its lack of movement. The subtalar joint (also called the talocalcaneal or inferior ankle joint) exists between the talus and the calcaneus. The midtarsal joint (also called the transverse tarsal joint) divides the foot into anterior and posterior portions. On the proximal side of the joint are the head of the talus and the end of the calcaneus. Distally are the navicular and the cuboid. All of these joints are linked together so that when one moves, both are affected. The talonavicular articulation has the form of a ball-and-socket joint with a good degree of freedon,. The calcaneocuboid joint is also somewhat free. However, these two parts of the midtarsal joint constrain one another. When both articulations are aligned with one another, a small degree of flexion/extension is permitted in the midfoot enough to flatten the arch. When the calcaneus is inverted under the talus, the former axes are brought out of alignment. No movement may occur and the foot is rigid. In addition, inversion of the calcaneus twists and tightens ligaments crossing the midtarsal joint, further stabilizing it. The foot is therefore able to shift back and forth between mobile and immobile positions. This design of the joints, plus the muscles and ligaments that prevent the bones from separating may be referred to as the conversion mechanism of the foot (Mann et al, 1964). The joints around the talus, subtalar and transverse tarsal, are critical in establishing the arch. The interactions of these creating the conversion mechanism are described above. The rigid supinated position of the joints corresponds to the highest position of the arch. Pronation loosens and lowers the arch. Normal variation in the design of these joints can affect the performance and health of the arch. The arch is strengthened by long tendons of muscles active during push-off in gait, the tibialis posterior, flexor hallucis longus, and flexor digitorum longus in the posterior compartment, and many of the smaller plantar intrinsic muscles. As these contract, they pull the bones more tightly together, especially across the medial side of the transverse tarsal joint; and they resist passive dislocation of the joints, (Morton, 1924; Mann et al , 1964; Ankel-Simons, 1983).

Skeletal Maturation

The process of bone formation (osteogenesis) involves three main steps: production of the extracellular organic matrix (osteoid); mineralization of the matrix to form bone; and bone remodeling by resorption and reformation. The cellular activities of osteoblasts, osteocytes, and osteoclasts are essential to the process. Osteoblasts synthesize the collagenous precursors of bone matrix and also regulate its mineralization. As the process of bone formation progresses, the osteoblasts come to lie in tiny spaces (lacunae) within the surrounding mineralized matrix and are then called osteocytes. The cell processes of osteocytes occupy minute canals (canaliculi) which permit the circulation of tissue fluids. To meet the requirements of skeletal growth and mechanical function, bone undergoes dynamic remodeling by a coupled process of bone resorption by osteoclasts and reformation by osteoblasts. Osteoblasts are derived from mesenchymal stem cells of the bone marrow stroma. They possess a single nucleus, have a shape that varies from flat to plump, reflecting their level of cellular activity, and in later stages of maturity line up along bone-forming surfaces. Osteoblasts synthesize and lay down precursors of collagen 1, which comprises 90-95% of the organic matrix of bone. Osteoblasts also produce osteocalcin (the most abundant non-collagenous protein of bone matrix) and the proteoglycans of ground substance and are rich in alkaline phosphatase, an organic phosphate-splitting enzyme.Osteoblasts have receptors for parathyroid hormone and apparently for estrogen.Hormones, growth factors, physical activity, and other stimuli act mainly through oteoblasts to bring about their effects on bone. The collagen 1 formed by osteoblasts is typically deposited in parallel or concentric layers to produce mature (lamellar) bone. But when bone is rapidly formed, as in the fetus or certain pathological conditions (fracture callus, fibrous dysplasia, hyperparathyroidism), the collagen is not deposited in a parallel array but in a basket-like weave and is called woven, immature, or primitive bone. In fully decalcified bone sections, the extracellular matrix stains pink with H+E, similar to collagen elsewhere but with a more homogeneous than fibrillar structure which latter is easily observed by polarizing microscopy. The main mineral component of bone is an imperfectly crystalline hydroxyapatite $[Ca10(PO4)6(OII)2]$ which comprises about 1/4 the volume and 1/2 the mass of normal adult bone. The mineral crystals, as shown by electron microscopy, are deposited along, and in close relation to, the bone collagen fibrils. Calcium and phosphorus (Pi, inorganic phosphate) are, of course, derived from the blood plasma and ultimately from nutritional sources. Vitamin D metabolites and parathormone (PTH) are important mediators of calcium regulation, and lack of the former or excess of the latter leads to bone mineral depletion. The extracellular matrix of bone is mineralized soon after its deposition, but a very thin layer of unmineralized matrix is seen on the bone surface, and this is called the osteoid layer or osteoid seam. In some pathological conditions, the thickness and extent of the osteoid layer may be increased (hyperosteoidosis) or decreased. Hyperosteoidosis may be caused by conditions of delayed bone mineralization (as in osteomalacia/rickets resulting from vitamin D deficiency) or of increased bone formation . Osteoclasts are derived from hematopoietic stem cells that also give rise to monocytes and macrophages. Typically multinucleated, osteoclasts adhere to the surface of bone undergoing resorption and lie in depressions termed Howship's lacunae or resorption bays. The boundary between the old and new bone is distinguished in an H+E section by a blue (basophilic) line called a cement line or reversal line. Osteoclasts are apparently activated by "signals" from osteoblasts. For example, osteoblasts have receptors for PTH whereas osteoclasts do not, and PTH-induced osteoclastic bone resorption is said not to occur in the absence of osteoblasts, (Roche et al,1996; Ankel-Simons, 1983, 1999; Marieb, 1997,2000).

At an early stage of embryonic development, a cartilage model of much of the skeleton of extremities, trunk, and base of the skull) is formed from the mesenchyme, (each pathway differing per species). In the further fetal development of long bones, a rim of primitive bone is first laid down in layers over the middle of the shaft by osteoblasts arising from the overlying periosteum, and subperiosteal bone formed in this way soon extends up and down the shaft (diaphysis). The process by which bone tissue replaces membranous fibrous tissue is called intramembranous ossification. The bones of the limbs and vertebral column are endochondral (they are first formed as cartilage which is gradually replaced by bone). Centers of ossification in the cartilage gradually enlarge to form the bone. The shaft of long bones is referred to as the diaphysis, or primary center of ossification. The ends of the long bones have secondary centers of ossification, the epiphysis, that are separated from the diaphysis by cartilage plates. Eventually, the cartilage plates, the epiphysial cartilage, are replaced by bone and the epiphyses unite with the diaphysis to form a single bone. The flat bones of the skull, mandible, and clavicles are intramembranous bones, formed from membranes, and generally do not have epiphyses. The skull itself has a separate growth center, the neurocranium. This is the braincase itself, composed of cartilage and /or endochondral bone, (Ankel-Simons, 1983; Marieb, 1997;2000).

Locomotor Anatomy

Much of primate anatomy reflects habits of movement and commonly utilized substrate. Since the powerful legs of most primates are slightly longer than their arms, the pelvis is normally higher than the head when standing quadrupedally. However, a few species exhibit extreme locomotor specialization, emphasizing arms for arm-swinging, legs for leaping, or arms and legs comparable in length for quadrupedal climbing or walking on the ground, behaviors which are directly correlated both with searching and obtaining food and the postures assumed while doing so. Therefore, limb length can serve as an index of the relative emphasis upon arm versus leg for movement, (Swindler et al, 1973).

Short limbs with legs and arms of comparable size are found in those who are habitually quadrupedal and arboreal. Most of the time there is a good amount of sexual dimorphism among primate species, particularly in those who are arboreal. Females have an easier time maintaining movment in the trees due to the fact the limbs are shorter and they are lower to the ground with a lower center of gravity making balancing easier than for males. This is also true when comparing larger and smaller arboreal primates overall. Therefore, having shorter limbs tends to be an advantage to arboreal life, making it quicker and easier to move and balance in order to obtain food or escape predation. These animals walk on longer horizontal tree branches and they are normally shorter and more robust than other primates. Other characteristics include moderately sized fingers and toes, prehensile hands and feet, and somewhat mobile shoulder joints located and directed sternally on the thorax. Some species, especially those who can leap, have flexible and elongated backs with extra vertebrae and powerful muscles associated with both the back and hind limb. Some prehensile tails also serve to anchor the animals to branches, (Swindler et al, 1973; Herskovitz, 1977;Fleagle et al, 1999).

Longer limbs with legs and arms of equal length are found in those quadrapeds which are terrestrial. These primates tend to have shortened digits and elongated, robust tarsal and metatarsal elements. The shoulder joint, lying alongside the narrow and laterally flattened thorax, is oriented toward the ground. The weak clavicular-sternal joint is easily dislocated from the segmented sternum. Arms and legs, modified for powerful sagittal motions, have a relatively small range of motions. The humerus has a prominent deltoid process, the insertion point of the M. triceps brachii, a powerful forearm extensor. The hands are pronated when in contact with the ground. One species, the patas monkey, is digitigrade (like a cat) , with specialized anatomy for terrestrial running in which only the fingers and toes make contact with the ground, (Kavanaugh, 1983, Ankel-Simons, 1983,1999). The vertebral column is the axis of the body's support for all parts of the skeletal system. All are protected by segments of bone and cartilage, and areas of specialization are grouped under the names cervical, thoracic, lumbar, sacral, and coccygeal, according to the regions they are located in or associated with. The total number of vertebrate a particular species possesses depends upon the type of locomotor behaviors it practices.So, the number is sometimes increased by an additional vertebra in one region, (particualrly with respect to the presence of a tail) or it may be diminished in one region, the deficiency often being supplied by an additional vertebra in another.The vertebra in the upper three regions of the column remain distinct throughout life, and are known as true or movable vertebra, while those of the sacral and coccygeal regions are referred toas false or fixed vertebra, because they are united with one another in the adult to form two bones, which are found forming the upper bone or sacrum, and the terminal bone or coccyx, (Ankel-Simons, 1983; Aiello, 1990; Marieb, 1997;2000).

In suspensory primates, the bones are under tensile rather than compressive forces. So, the bones can be thinner due to reduced bending stress. Their trunks are Relatively short and broad. Short lumbar region - decreases flexion and extension during hanging and reaching. The scapula is located dorsally rather than laterally , which increases range of movement, thereby also decreasing tensile stress. Arboreal, as well as terrestrial, quadrupeds have extremely well developed clavicles which act as struts to aid in the resistance of compressive forces. Leapers have longer lumbar regions for both increased flexion and extension, making movement faster while decreasing overall stress. In bipeds, there is dual curvature in the spine, which is designed with vertebrate supported by cartilgeneous joint articulations for absorption of compressive stress. The curvature design moves centre of mass forward and closer to hip joint to aid balance as well as to decrease stress, since the vertebrae increase in size from the cervical region to the lumbar increasing weight-bearing capability, (Kavanaugh, 1983, Aiello, 1990;Ankel-Simons, 1983,1999). Very long limbs with legs and arms of comparable length on quadrapedes who are arboreal with an emphasis on quadrupedal climbing and suspension. A few primate species in the Family Lorisidae combine quadrupedal suspensory climbing with quadrupedal arborealism, requiring great joint mobility and a wide range of movements. Their hands and feet are particularly prehensile. (Napier et al, 1967).   Arms longer than legs occur in brachiators who are arboreal. Brachiation is a special form of locomotion in which the body is suspended below the branches. It allows utilization of small branches near the fringes of a tress canopy since the brachiator is suspended beneath them. In contrast, large bodied quadrupeds that try to walk on small branches have difficulty in balancing as the supporting tree limbs bend. A brachiator can easily exploit such areas of tree canopies by dispersing its weight to the ends of several branches. New World brachiators use their prehensile tails as fifth limbs to further disperse weight. Most rapid brachiators use gravity to convert vertical height to speed., (Swindler et al, 1973;Herskovitz, 1977). Brachiation is generally associated with major alterations in the arm, head, hand and thorax. The shoulder joint is positioned laterally and cranially on a barrel shaped thorax. Robust muscles attach to the sternum, vertebral column ,head, and rib cage, stabilizing the shoulder. The more powerful the arm movements, the more robust the muscles are. The clavicle acts as a strut to stabilize the shoulder joint against a sternum whose segments join to form a single bone. This clavicular-sternal joint is strong and is not easily dislocated. A relatively round head of the humerus indicates a very wide range of motion. Additional elbow strength results from a more distinct seperation of the radius and ulna on the articular surface of the distal humerus. The olecranon process of the ulna is small, allowing full extension of robust forearms. Brachiators tend to have reduced thumbs, and if a thumb is even present, it is folded out of the way against the palm where it does not interfere with elongated fingers that hook to branches. The lumbar region of the vertebral column is shortened and stabilized, and a very mobile hip joint allows the foot to grasp on and anchor in a range of positions, (Hill, 1974; Kavanaugh, 1983, Ankel-Simons, 1983,1999).

In terms of locomotor types , Napier and Davis (1959) were the first people to actually define the word 'brachiator' as an arm-swinging primate. However, the earliest use of the word can be seen in Owens work back in 1859 when he described it as 'locomotion by means of the arms in the manner of gibbons'. Brachiation is now thought of as a highly specialised form of suspensory locomotion which can be broken down into five different types: True brachiation as demonstrated by gibbons involves the use of the pectoral limbs to support the body in suspension beneath a superstratum as the animal moves in a swinging. Modified brachiation as seen with the Great Apes. Great Apes are not thought to brachiate but they seem to have a morphology that suggests they evolved as brachiators. Pro-brachiation as seen in fossil pongids such as Proconsul (a Miocene African Hominoid). New World semi-brachiation, which is seen in the prehensile-tailed cebids or Ateles (Spider monkey). These animals are more or less brachiators today but have achieved this position independently of gibbons. Old World semi-brachiation as in Presbytis (Asian Leaf monkey, langur).The only true brachiators are gibbons but even they do not totally rely on brachiation to get them around their territory. Fox (1972) and Elfson (1967) both broke down gibbon locomotion into five main types: Brachiation, which involves travel along a branch via a series of swings below alternating handholds. Climbing, which involves any continuous progression using three or more limbs. Bipedalism, which involves walking on the two hind legs with the forelimbs abducted atthe shoulder and used for balance.Leaping or diving , which involves the gibbon jumping or dropping from a higher to a lower level. Finally, there is swinging, or arm swinging, which involves the gibbon swinging from one branch or bough with one or both hands in contact with the handhold. Chimpanzees, gorillas and humans are capable of this type of brachiation, but do not practice it habitually as a means of locomotion. The orangutan combines quadrapedal climbing with brachiation, but like chimps and gorillas, it habitually a terrestrial quadruped, (Hill, 1974; Herzkovitz, 1977; Kavanaugh, 1983, Aiello, 1990; Ankel-Simons, 1983,1999).

A higher forelimb/hindlimb ratio is seen in habitual quadrupeds that knuckle-walk or fist walk. This is quadrupedal locomotion in which the hands are pronated and fingers flexed resulting in dorsal surfaces of the middle phalanges contacting the ground, supporting the weight on the knuckles. Gorillas and chimpanzees are habitual knuckle -walkers, whereas orangutans usually move quadrupedally with the hand made into a fist, (Herskovitz, 1977). On the other hand, a higher hindlimb/forelimb ratio is found in those primates who are leapers and are arboreal. A special class of leaping locomotor behavior , in which the body is positioned vertically at rest, is called vertical clinging and leaping. It requires powerful hind limbs to propel the leap as well as to break the impact of landing. Most, but not all, vertical clinging and leaping species have a tail that is used to maintain attitude control during leaps. Rapid movements are well coordinated during the leap that the animals transits the crown of a tree without appearing to ever have made contact with the branches, (Herskovitz, 1977; Ankel-Simons, 1983). There is a tendency towards elongation of tarsal elements, especially calcaneus and navicular. Posterior elongation of tuberosity of the calcaneus serves as a robust lever arm for M. gastrocnemius and M. soleus muscles, powerful flexors of the foot. The tendency for fusion of the tibia and fibula is fully expressed only in the Tarsiers, (Napier et al, 1967;Herskovitz, 1977, Ankel-Simons, 1983).

Legs longer than arms are found in those primates who are bipedal. This is only found among humans, though many other primates are capable of facultative bipedalism. Foot specializations for bipedalism include an enlarged and robust tarsal region, greatly reduced phalanges, and strong ligaments that bind tarsals and metatarsals into shock -absorbing longitudinal and transverse plantar arches. A large calcaneus tuberosity acts as a lever arm for plantar flexion. The most unique characteristic of the long, robust legs is the placement of the knees close to the median sagittal plane, functionally beneath the body's center of gravity. The knee itself is adapted to locking in full extension with deep grooves to stabilize the patella, a bone that forms in tendons of the quadriceps muscle. The broadened hip becomes a weight-bearing joint, characterized by an enlarged femur head as a weight -bearing bone surface. Pelvic anatomy is dramatically rearranged. A relatively broad sacrum positioned above the hip joint transfers weight to the femur head through the wide and more robust illium. A shortened ischium places the ischial tuberosity relatively close the the acetabulum. The vertebrae, increasing in size progressively from skull to sacrum , are arranged in a ventral-dorsal s-shaped curve above the pelvis. Though free of locomotor function, the arm retains the range of motion seen in brachiators. The small, light, yet highly specialized dentition found among vertical clingers and leapers tends to aid in the attitude control of the body. Since the tooth comb is designed for highly specialized grooming, it acts to keep the tail clean and free of debris that could weigh it down and cause the fur to matte up, thereby decreasing attitude efficiency, since their tails act as rudders while they are in motion. (Hill, 1972; Swindler et al, 1973; Kavanaugh, 1983;Roche et al,1996;Marieb, 1997, Ankel-Simons, 1983, 1999).

There is also one additional method of primate locomotion, climbing by nails. Elongated and laterally compressed nails of callithricines have the functional attributes of claws. Although they climb by grasping small branches, they are able to use these specialized nails to cling to relatively flat , vertical ,surfaces of larger trees, (Herskovitz, 1977).

Primate Vision

Color vision is a primate characteristic that presumably reflects an arboreal ancestry. It is evident in those displaying diurnal lifestyles, and it is thought to have evolved as a result of the greater reliance on sight and the need for distinction of objects. Though all primates are visually oriented, not all posses color distinction. The eyes are structured in order to point reflect the areas of light perception, or their retinas. The innermost layer of the eye is the retina, centrally placed. This is the area where optical impulses are received, and a direct connection exists with the optic nerve. The retina is composed of an outer pigmented region with an inner layer of specialized nerve cells. These cells are receptors, they are cylindrical rods arranged perpendicualr to the retinal surface. On the outer rims of these rods there is rhodopsin (visually, purple in color). This is a specialized pigment which facilitates the absorption of low intensity light waves, making the rods sensitive to the differences between black and white. Rods are predominantly invovled in night vision, therefore, nocturnal primates primarily have rods associated with their vision abilities. The second type of specialized cells are the cones, which are just that, conical in shape. The broader end is facing toward the center of the eye, and they contain the pigment iodopsin. The cones act to provide visual color activity abilities, abilities which are common among diurnal primates. It is among the higher primates that the retina is found to be composed of both types of light receptors. Though it has been found that prosimians also posess color vision abilities, yet to a lesser extent than anthropoids. Meaning, color vision cones are more developed in the higher rather than lowe primates, and although the eyes of lower primates are quite forward-facing, the left and right visual fields do not overlap as much as they do in higher primates. This arrangement limits depth perception to the central part of the field of view. And whereas it is natural that the nocturnal lower primates lack color-sensitive cone cells in their retinas, what little is known about visual discrimination in the diurnal lemurs suggests that their color vision is at best limited, (Ankel-Simons, 1983;Schnapf et al, 1999; Schneeweis et al, 1999).

The above mentioned photoreceptors, the rods and cones within the retina function at varied intensities of light. The rods function better at low light levels (scoptic vision) , while the cones respond to much higher light intensities (photopic vision). The eyes of most diurnal mammals have cones more numerous toward the center of the retina (the region of sharp focus) and more rods toward the periphery. Nocturnal primates have only rod photoreceptors in the retina, though some have very few cones as well. The retina of higher primates has a macula lutea (yellow spot) of cones. The focea, a small depression in the center of the macula in which there is only a single layer of cones, is the area of keenest vision and the target of focusing by the lens. The color perception allowed by the cones is dependent upon the relative degree to which the rhodposin and iodopsin are stimulated by varied light intensities which act to breakdown these receptors inorder to generate an action which generates nerve impulses to the brain in order to create vision. Primates have thress different pigments, producing trichromatic vision, (Hill, 1972; Savage, 1972, et al, 1986;Marieb, 1997;Fleagle 1988;Schnapf et al, 1999).

Anthropoid vision is stereoscopic. Their eyes are positioned forward, allowing and overlap of most of the fields of vision with the optic axes parallel. An object is focused on both retinas simultaneously. The optic nerve tracts that pass information from the retina to the brain meet at the optic chiasma. In most vertebrates the fibers of the optic nerves cross at the chiasma and pass to the opposite of the brain. However, in mammals, some of the fibers do not cross over, so information from each eye is processed in both hemispheres of the brain, (Hill, 1972; Savage et al, 1986;Marieb, 1997;Flea gle 1988).

Overview of Primates and Primate Identification

There are approximately 680 primate species overall, 310 extant, (Napier et al, 1967;Kavanaugh, 1983). Primates range in size from the 160 kg male mountain gorilla to the less than 100 gram pygmy marmoset. Primates evolved from insectivores (tree shrews, or tupaias, were once classified as the most primitive primates. Now they are considered a separate order by some, while others see them as divergent lineages). The general evolutionary trend within the order is towards stereoscopic color vision (shortening of the muzzle, flattening of the face, decline in the importance of smell) and refinement of hands and feet as grasping organs (flat nails instead of claws, sensitive pads for gripping). This evolution has culminated in the versatile human hand with its completely opposable thumb. Other trends are an increase in brain size; a reduced reproductive rate (many primates give birth to single offspring), and an increase in social complexity. There are specialized limbs, eyes, noses, brains, teeth , all identifiable and distinguishable based upon analysis of skeletal features and skeletal analysis. Primates can be distinguished not only between one another, but also in terms of their ancestral affinities, divergence times , and in the case of humans, racial categories (or geographic regions of origin). Each has demonstrated adaptive features to individual environments, which has been the primary determinant in altering skeletal features, (LeGros Clark, 1961; Ankel-Simons, 1983, Rowe, 1996; Fleagle 1988). gle 1988).

Applicable Identification Techniques

Estimating the Age of the Individual

During the period of growth and development, age estimation is usually quite precise du to evidence available through growth plates and centers throughout the body. However, with the cessation of growth and eruption of the permanent dentition, assessment of age becomes more difficult, relying on evidence of degenerative processes. Age estimations in adults are therefore made from such markers as dental eruption, dental attrition and dental loss , which are indicators of older age categories. Severe tooth wear is common in prehistoric individuals, and can sometimes prove useful if there is some indication as to the types of diets they had (this could be indicated in the types of teeth and both their cusp and overall functional structures as well though). Skeletal maturation of epiphyses to diaphyses is a fairly accurate guide to age. Though cranial suture closure is more variable, and it is useful for older age ranges. Many areas of the older adult skeleton undergo progressive deterioration that serves as useful indicators of age. The pubis symphysis and sternal ends of ribs are particularly useful. As living bones age, there is a slight but cumulative loss of osteons and a greater degree of ossification (the circular structures of bone surrounding the haversian canal), (Bennett, 1993; Reichs,1998).

In terms of the pubic symphysis, Morphological changes of the articular surfaces of the innominate provided the best postcranial age indicators. Variation in the face of the pubic symphysis, the anterior-most point of articulation between the two innominates in the pelvic girdle, is a common region analyzed for age determination. Changes in the symphyseal surface over time proceed in a predictable pattern from a heavily contoured face, to one delimited by a rim in the mid 30s, to a surface marked by increasing porosity after 40 years, (Brothwell, 1981; Bennet, 1993; Reichs, 1998).

The degree of tooth wear (attrition), when seriated within a population, is a useful indicator of an individual's age at death. Again, dental attrition varies with types of food consumed. So, again, it is important to compare teeth within the group of interest, and understand their mode of subsistence. Prior to any estimations based upon toothwear, it is important to have an inventory of the teeth present and their general degree of preservation. Primates display reduced complexity in the heterodont dentition, a characteristic most notable in the omnivorous mouths of humans. Food is taken into the mouth with the anterior dentition, the incisors and canines, and processed by the premolars and molars. These posterior teeth grind, pulp, slice and dice the foods consumed, which increases the surface area for the digestive enzymes in the mouth to act upon to begin the digestion process, (Brothwell, 1981; Hilson, 1996; Miles, 1962; White, 1999).

The normal aging process can also adversely affect tissue properties leading to degenerative conditions which affect tendons, ligaments, articular cartilage and bone. Articular cartilage is a tissue which allows near frictionless movement between bones yet transmits high loads across the joint. As a result, it thins over time, and is evidnet in conditions such as osteoarthritis, which occurs naturally over time as an idividual ages due to this thinning in conjuntion with subchondral bone density changes, and mechanical property changes along with ossification within the tendons and cartilage as well,(Brothwell, 1981; Bennet, 1993; Reichs, 1998).

Estimating gender

Gender can be distinguished on the basis of size and robusticity. Males are 5-10 % larger and heavier (in humans, though non-human primate males also tend to be larger than females overall). Females as a group tend to be smaller and have finer muscle markings. However, when dealing with a specimen of an unknown background it is difficult to determine just what is large or robust. It is also difficult to make accurate gender estimates on immature specimens since many distinguishing characteristics are evident only following adolescence, (Rowe, 1996;Reichs, 1998).

The most dramatic differences between males and females are in the pelvis where sexual dimorphism reflects the compromise between childbirth and locomotion. Since the heads of the humerus and femur reflect body size, the diameter of these two features provides a guide: less than 41mm for females, and greater than 45mm for males. The gender of the pelvis is recognizable in most individuals by the ration of the length of the ischium divided by the length of the pelvis, and a value then of greater than .9 indicates female, (again, the values represent markers for humans and not non-human primates),(Lohman et al, 1988).

A similar relationship works also for the foot. If the maximum length of the os talus is greater than 52mm , the individual is likely to be male. The skull is often used as well as a guide to gender. Males tend to have larger and more robust faces, especially the supraorbital torus. If the upper border of the orbit has a knife-like margin, the individual is likely to be female. A mastoid process that is large is also a male indicator. The male mandible tends to be larger, with an outward flaring gonial angle (approximately 90 degrees) and a square chin. A female chin is more pointed, (Rowe,1996;Reichs, 1998).

Estimating Ancestry

A useful technique in estimating the gender or biological ancestry has been to examine cadavers of known age and background to develop measurements and computations that would correctly distinguish gender and population (unclaimed bodies from morgues or bodies donated for medical dissections). Since it is questionable that most such samples are comparable to present populations, such statistics are used only as a supplement or aid in the estimation, (Kennedy et al, 1992;Saur, 1992 Reichs, 1998).

There are certain features which are more common in one group than in other groups. These features are mostly present in the facial region. A point to be made, however, is that there is more individual variation within races than variation between races. There are certain traits that tend to be more common in people of specific ancestry. One such trait is the shape of the incisors; people of African ancestry more commonly have blade shaped incisors, Australian ancestry tend toward trace shape, East Asian ancestry often have shoveled incisors and European ancestry rarely have shoveled incisors, (Groves, 1989). Skulls tend to be the most reliable bones of evidence, particularly since they exhibit racial differences in higher or lower zygomatic bones/arches, rounded or more oval orbits, wider versus more narrow nasal passages, and smoother versus a more rough contour to the chin. Though Asian and European populations vary little between one another, such features can be distinguished enough from a skull to at least indicate such geographic regions as points of origin or ancestry, (Andrews et al, 1976; Molnar, 1998).

Discriminate function analysis is used to determine which variables discriminate between two or more naturally occurring groups, or in the case of remains identification, between two or more features in order to establish a better idea of race, though it also has been used to aid in identifying gender or age. It involves the taking of various measurements to the nearest millimeter. Each measurement is then multiplied by a coefficient specified in the functional formula, and the products are added or subtracted. If the sum is above the value specified as a sectioning point, the estimate then indicates membership in the group specified by that equation, (Kennedy et al, 1992;Saur, 1992 Reichs, 1998). Also, racial differences in bone mineral density, as well as overall bone density measurements have allowed for the distinction of placement of individual specimens into specific geographic areas of origin. This, along with identification of distinguishing craniofacial features, allows race, or ancestral affinities to be estimated, (Saur, 1992 ; Reichs, 1998; Nelson, 2001).

Conclusion

The fossil record of the primates is exceedingly diverse. It depicts isolations, geographic adaptations, radiations and emergences, lineage/phyletic histories, and brings together a living picture of the relationship between the many species of extinct primates, the earth and geological time, and the extant species of the present. Regardless of what the fossil record may be lacking, it still provides the best information pertaining to diversification and adaptation from the past into the present, and allows a glimpse into how primates can be identified and related in terms of their evolutionary trends based upon morphological, geological and molecular data, leading to an understanding of such relationships and identification markers known throughout extant primates today, both human and non-human. The fossil record is still fairly restrictive in terms of its own geographical range however, and parts of Africa, South and Central America, along with

Southeast Asia still await further exploration.

Data today is still collected and interpretations (inferences) are still made on the basis of comparative morphologies and the reconstruction of ecological diversification utilizing faunal remains of any given location in conjunction with what molecular data may be available from both extant and extinct populations. This offers some reconstructive potential then of the types of relationships which occurred and the types of adaptations which took place among the varied genera and species giving some indication of how they were either similar or different from those same things used to characterize present day radiations. Body size and dentition are known to be directly correlated to such things as the econiche a particular species inhabited, diet and locomotion habits. Each feature affects the other. Faunas provide then some indication of diversity and geographic distribution, and some insight into why and how some particular adaptations occurred as well as the reasons for varied degrees of sexual dimorphisms and the retention or not of ancestral (primitive) traits in some species.

Evolutionary patterns can be seen in total skeletal morphologies in several areas. The first being size. There has been a good amount of diversity in terms of size given the range of climates and conditions the varied species of primates inhabited over time and space. Primates, as a whole, have increased in size over the past 65 million years when they first began to appear and diversify. Many of the known features were early adaptations which occurred more rapidly during the Paleocene and Eocene when primates were first on the scene, and they were slower to occur over the Oligocene through the present time (Holocene), (Fleagle 1978; Gibbons, 1981; Covert, 1986). Many of the newer species, human lineages and apes in particular, did not begin to appear until the Miocene and are not found in earlier times, but can be traced on the basis of total morphological features, molecular evidence, behavioral patterns along with evidence of dietary diversity, back to the first primates of 65 million years ago, (Le Gros Clark, 1959; Fleagle et al, 1985).

Dietary adaptations are directly tied to ecological diversity as well, in that over the course of primate evolution, there was considerable variation depending upon a particular species' niche, and at that, even a frugivorous plesiadapid was somewhat different anatomically and behaviorally from a frugivorous adapid , particularly with respect to dentition, (Covert, 1986; Fleagle, 1988). Foraging strategies would differ as well. Old World anthropoids have changed a good deal in terms of their own diversity, as have the Early Oligocene higher primates which are clearly shown throughout the Fayum deposits of Egypt showing they were primarily frugivorous anthropoids all based upon changes in dentition over time, (Kay, 1997). Later, during the Miocne, higher primates with teeth showed more of a folivorous habit and began to appear more like anatomically modern apes and humans in terms of dentition and adaptive morphologies with respect to such genera as *Dryopithecus, Oreopithecus*, and *Ramapithecus*, leading later to the pongids and hominids, (Brown *et al*, 1985; Fleagle , 1999).

Again , the lack of adequacy in the fossil record an make it difficult to also note all temporal changes in locomotor habits, but overall patterns can be detected nonetheless. In the plesiadapiformes the ankle and claws alone are indicative of an arboreal lifestyle, though they lack the opposable thumb and longer and more curved claws associated with arborealism of the Eocene through the present (Holocene). So, this shows they were arboreal , but most likely did not leap the way some extant species (tariers for example) today do. For comparison and to demonstrate changes, it is interesting to note that the Eocene fossil prosimian species are similar to the extants in that they still have maintained a more generalized anatomy. For the most part, they have remained arboreal quadrupeds and quadrupedal leapers (cheirogaleids for example) , while in some, a more specilaized vertical clinging habit is seen as is the case among extant indriids and tarsiers, (Covert, 1995; Fleagle *et al,* 1997).

Phylogenies , again, are depicted through the fossil record and inferred on the basis of morphological and molecular data, (Delson, 1984; Howells, 1997). In the case when there are few fossils and they are known from a wide range of time, then it is thought that they are representative then of a series of intermediate or transitional lineage forms stemming from one species. These relationships become clearer, or can change, as more fossils become known and more genera, species , etc are defined or even redefined on the basis of analyis of new materials. Though in general terms, the initial appearance of the majority of the taxa is known by the rapid proliferation of forms, which brings about an evolutionary tree looking like a branching bush (ie: new buds and lineages of branches stemming from the same base), and it is through the finding and identification of new fossils and new faunal evidence that the tree is formed and or altered, affecting how speciation and relatedness are perceived, (Simpson, 1945, 1949; LeGros Clark, 1959; Gibbons, 1981; Carroll, 1997).

New species, afterall, are only modifications of the previously known forms which have adapted to new econiches throughout time.   What emerges are overlapping traits and total morphological patterns consistent with the ideas of intermediate forms displaying mosaic evolutionary trends. Mosaic evolution is considered a pattern of the adaptiveness of certain anatomical traits throughout the lineages of the many species which existed in comparison to what is then seen in the anatomical features of extant species over the course of the entire history of their taxonomy (genera, families, etc). What is seen in the extant species is the result of competitive econiche and climatic changes in response to challenges presented by food supplies, types of food stuffs, competition with other species,predation, geological and climatic changes, and overall habitat changes leading to adaptations which then become evident in the fossil record as the previous species become extinct and new intermediates form continually becoming those which are considered the living species at any given time, (Darwin, 1859;McHenry, 1975; Gibbons, 1981;Wolpoff, 1984;Bown *et al*, 1987;Carroll, 1997).

Certainly, the origin of the earth is important, as is the first life appearing on it. Following that , there is the origin of complex single celled organisms (protozoa), then multi-celled organisms (metazoa), animals with backbones (vertebrates), first land vertebrates, and even mammals. Traditionally, however, there is a limit to the study of human evolution (paleoanthropology), which is the study of just one mammalian order: that of the primates (Primates).The aim of this paper has been to discuss what a primate is, and provide an introduction to the range of primates present in the world today, their histories, and how both non-human and subsequently human primates evolved through time for the purposes of establishing an appreciation of the wide variation in these animals which is essential for any discussions pertaining to how humans evolved. Like many definitions, the definition of what makes a primate (as opposed to a rodent, or a carnivore etc.) is complex. There is little argument as to the core groups of animals today that are primates asas has been illustrated throughout this paper, but as one goes back in the fossil record, there is more dissension. LeGros Clark (1959) pointed out continually the need to observe the total morphological patterns , and Simpson (1945, 1949) stressed the need to combine morphological data with molecular (or protein data at the time) in order to establish phylogenies. Mirvart (1873), of course , provided a much needed and still highly applicable and purely descriptive definition as a starting point in the process of understanding primate evolution, phylogeny and taxonomy: "Unguiculate, claviculate, placental mammals, with orbits encircled by bone; three kinds of teeth, at least at one time of life; brain always with a posterior lobe and calcarine fissure; the innermost digit of at least one pair of extremities opposable;

hallux with a flat nail or none; a well developed caecum; penis pendulous; testes scrotal; always two pectoral mammae". Though there is no characteristic feature of the order primates that every primate shares or that does not exist in other orders of animal. This makes recognition of fossils and determination of evolution difficult. For example, the question of when and why primates evolved is impossible to answer if the exact characters that constitute a primate cannot be defined. The definition of a primate is therefore, of key importance in the study of primate origins and primate taxonomy in conjunction with morphological and molecular remains.

The primates are divided into a basic two groups, moderns (extants) and fossil. The best way to look at such a diverse and complex group throughout geological time would be through the utilization of the pattern concepts, or examining the total morphological patterns occurring over time as well as the evolutionary patterns (or trends) found throughout the order. No one species is defined or classified on the basis of one or even a few select characteristics, but in the general patterns displayed in both human and non -human primates. Taxonomy was originally designed to make it easier to understand these evolutionary trends and relationships between lineages leading to extant primates as well as humans, but it has become one of the most complex games of paleoanthropology with oversplitting, overlumping, due to the issues of gaps in the fossil record. Though, it is still based on total morphological patterns, and despite its complexities and controversies , it is still one of the most interesting components of primate evolutionary study. Many controversies have been around for centuries, some resolved, others not, and new ones emerge continually. Nonetheless, new molecular and morphological evidence continues to surface which will ensure a never ending quest for piecing together primate evolutionary history , primate phylogeny, and the particularly interesting aspects as to where human primates fit in the total evolutionary scheme. Integration is the key to this, as many factors have had profound affects up the appearance and adaptive features of all primates, as well as being the root of confusion and causes of overlumping or splitting throughout all of primate (human in partiuclar) evolution. Such factors which must be taken into consideration at all times are: climatic events, faunal remains, geological events and time, econiche alterations resulting in ring speices formations, biological and physiological responses to environmental stressors, and behavioral adaptations which also occur in relationship to all environmental changes. If such factors are not integrated into the evolutionary scheme overall, the beauty of the process is not only lost, but skewed for all of evolutionary history, (Darwin, 1859;Huxley, 1872;Clark, 1926,1934, 1959; McHenry, 1975; Gibbons, 1981;Wolpoff, 1984;Bown, 1976, *et al*, 1987;Carroll, 1997).

Marston Bates perhaps summarizes this paper best with the following analogy: " I object to dividing the study of living processes into botany, zoology, and microbiology because by any such arrangement, the interrelations within the biological community get lost. Corals cannot be studied without reference to the algae that live with them; flowering plants without the insects that pollinate them; grasslands without the grazing animals". *Marston Bates- The Forest and The Sea, 1964.*

Appendices

1. Taxonomy of the Extant Primates and Geographic Ranges

From : *Daryl G. Frazetti Spring 2001 Based on MF Gibbons, Jr*

Kingdom: Animalia **Phylum:** Chordata **Class:** Mammalia **Order:** Primates

1.**FAMILY**: **Cheirogaleidae**

GENUS: Microcebus : (mouse lemurs) …range: w. and s. coast of

Madagascar

SPECIES: Microcebus murinus : lesser mouse lemur

Microcebus rufus   : russet mouse lemur

GENUS: Mirza (Microcebus) ….range: N-NW /SW coasts of Madagascar

SPECIES: Mirza coquerelli: Coquerel's dwarf lemur

GENUS: Chelrogaleus (dwarf lemurs)… range: west coast of Madagascar

SPECIES: Cheirogaleus major: greater dwarf lemur

Cheirogaleus medus: fat-tailed dwarf lemur

GENUS: Allocebus: … range: highland rainforests of Madagascar

SPECIES: Allocebus trichotis : hairy-eared lemur

GENUS: Phaner:… range: western coastal forests of Madagascar

SPECIES: Phaner furcifer: fork-marked lemur

2.FAMILY: Lemuridae

GENUS: Lemur:…range: tropical rainforests and dry thornbush of

Madagascar and Comoro Islands

SPECIES: Lemur catta: ring-tailed lemur

GENUS: Petterus:... range: west and east Madagascar

SPECIES: Petterus coronatus: crowned lemur

Petterus fulvus: brown lemur

Petterus macaco: black lemur

Petterus mongoz: mongoose lemur

Petterus rubriventer: red-bellied lemur

GENUS: Hapalemur:... range: bamboo and marshland areas of

Madagascar

SPECIES: Hapalemur aureus: golden lemur

Hapalemur griseus: grey gentle lemur

Hapalemur simus: broad -nosed gentle lemur

GENUS: Varecia:...range: eastern coastal rainforests of

Madagascar - Higher altitudes (upper levels)

SPECIES: Varecia variegata: ruffed lemur

GENUS: Lepilemur:... range: Western dry forests of North

Madagascar

SPECIES: Lepilemur mustelinus: weasel-lemur, (sportive)

Lepilemur ruficaudatus

Lepilemur septentrionalis

Lepilemur dorsalis

Lepilemur edwardsi

Lepilemur leucopus

Lepilemur microdon

3. FAMILY: Indriidae

GENUS: Avahi:…range: coastal rainforests of Madagascar

SPECIES: Avahi laniger: wolly lemur

GENUS: Propithecus:…range: mixed deciduous and evergreen

forests of NW Madagascar

SPECIES: Propithecus diadema: Diadem sifaka

Propithecus tattersalli

Propithecus verreauxi: Verreaux's sifika

GENUS: Indri:…range: eastern coastal rainforests Madagascar

SPECIES: Indri indri: Indri

4.FAMILY: Daubentoniidae

GENUS: Daubentonia:… (Aye-Ayes)… range: east coast and NW

forests Madagascar

SPECIES: Daubentonia madagascariensis: Aye-Aye

5.FAMILY: Lorisidae

GENUS: Loris:…range: India and Sri Lanka (trop.rain and

dry/semidry Deciduous forests)

SPECIES: Loris tardugradus : slender loris: range : swamps

India and Sri Lanka

GENUS: Nycticebus:…range: rainforests of SE Asia, Assam,

Burma, Thailand, Indochina,

Parts of Malaysia and East Indian Islands

SPECIES: Nycticebus coucang: slow loris

Nycticebus pygmaeus: pygmy low loris

GENUS: Perodicticus:…range: Benis, Burundi, Cameroon,

Gabon, Ghana,Guinea, Ivory Coast,

Kenya, Liberia, Nigeria, Rwanda, Sierra

Leone, Togo, Uganda, Zaire

SPECIES: Perodicticus potto: Potto gibbon

GENUS: Arctocebus: …range: Camaroon, Congo, Equitorial Guinea, Gabon, Nigeria (low dense forests and Swamps)

SPECIES: Arcticebus calabarensis

GENUS: Galago: …range: forested and bush regions of Africa south of The Sahara

SPECIES: Galago alleni: Allen's bushbaby

Galago granti: Grant's bushbaby

Galago moholi: Southern lesser bushbaby

Galago senegalensis: Northern lesser bushbaby

GENUS: Otolemur:…range: Ethiopia

SPECIES: Otolemur crassucaudatus

Otolemur garnettii

GENUS: Euoticus:…range: Camaroon, Congo, Equatorial Guinea, Gabon, Nigeria

SPECIES: Euoticus elecantus: western needle-clawed Bushbaby

Euoticus inustus: eastern needle-clawed

Bushbaby

GENUS: Galagoides:…range: African rainforests

SPECIES: Galgoides demidoff: demidoff's galago

Galagoides zanzibaricus: zanibar bushbaby

6. FAMILY: Tarsiidae

GENUS: Tarsius:…range: East Indies rainforests (

common on Samar,

Leyte, Bohor and Minanao)

SPECIES: Tarsius bancanus: western tarsier

Tarsius pumilus: pygmy tarsier

Tarsius syrichta: phillipine tarsier

7.FAMILY: Callithricidae

GENUS: Callithrix:…range: Upper Amazon of S.

America, W. Brazil, SE   Colombia, E.

Ecaudor, Peru and N. Bolivia

SPECIES: Callithrix argentata: silvery marmoset

Callithrix humeralifer: santarem

marmoset

Callithrix chrysoleuca

Callithrix jacchus: common marmoset

Callithrix flaviceps: white-eared marmoset

Callithrix penicillata: white-headed marmoset

GENUS: Cebuella: (pygmy marmoset): range: Rainforests of Brazil, Ecuador, Colombia,   Peru

SPECIES: Cebuella pygmaea: pygmy marmoset

GENUS: Saguinus: (tamarins): range: Lower rainforests of South America.

SPECIES: Saguinus bicolor: bare-faced tamarin

Saguinus martinsi: Martin's tamarin

Saguinus fuscicollis: saddle-backed tamarin

Saguinus imperator: emperor tamarin

Saguinus inustus: mottle-faced tamarin

Saguinus labiatus: white-lipped tamarin

Saguinus leucopus: white-footed tamarin

Saguinus midas: red-handed tamarin

Saguinus tamarin: Negro tamarin

Saguinus mystax: moustached tamarin

Saguinus nigricollis: black and red tamarin

Saguinus graellsi

Saguinus geoffroyi: Geoffroy's tamarin

Saguinus tripartitus

Saguinus fuscicollis

GENUS: Leontopithecus: (Goldens)... range: Evergreen and Broadleaf

forests. Brazil.

SPECIES: Leontopithecus chysomelas: golden-headed Tamarin

Leontropithecus chrysopygus: golden-rumped Tamarin

Leontropithecus rosali: golden lion tamarin

8. FAMILY: Cebidae: (capuchins)

GENUS: Cebus: range: deep rainforests of South America

SPECIES: Cebus albinfrons: brown pale-fronted capuchin

Cebus apella: black-capped capuchin

Cebus capucinus: white-throated capuchin

Cebus olivaceus:weeper capuchin

Cebus nigrivittatus

GENUS: Aotus: (night monkeys)...range: Central And South America

SPECIES: Aotus azarae: Southern night monkey

Aotus infulatus

Aotus miconax

Aotus nancymai

Aotus nigriceps

Aotus trivirgatus: Northern night monkey

Aotus brumbacki

Aotus lemurinus

Aotus vociferans

GENUS: Callicebus:...range: Bolivia and Brazil

SPECIES: Callicebus brunneus

Callicebus calligatus

Callicebus cinerascens

Callicebus cupreus

Callicebus donacophilus

Callicebus dubius

Callicebus hoffmannsi

Callicebus modestus

Callicebus moloch: dusty titi

Callicebus oenanthe

Callicebus olallae

GENUS: Samiri: (squirrel monkeys)...range: South America

SPECIES: Saimiri boliveiensis

Saimiri oerstedii: red-backed squirrel

monkey

Saimiri sciureus: common squirrel

monkey

Saimiri ustus

Saimiri vanzolinii

GENUS: Pithecia: (saki)…range: NW Brazil, Colombia, Ecuador, Peru

SPECIES: Pithecia aequatorialis: Equitorial saki

Pithecia albicans: white saki

SPECIES: Pithecia irrorata: bald-faced saki

Pithecia monachus: monk saki

Pithecia pithecia: white-faced saki

GENUS: Cacajao (uakari)…range: Brazil, E. Peru

SPECIES: Cacajao calvus: white uakari

Cacajao melanocephalus: black-headed uakari

Cacajao rubicundus: red uakari

Cacajao calvus

GENUS: Chiropotes: (saki)…range: Brazil

SPECIES: Chiropotes albinasus: white-bearded saki

Chiropotes satanas: black-bearded

GENUS: Alouatta: (howlers): …range: S. Mexico,
Central and South America

SPECIES: Alouatta belzebul: black and red
howler

Alouatta caraya: black howler

Alouatta fusca: brown howler

Alouatta palliata: mantled howler

Alouatta seniculus: red howler

Alouatta villosa: Guatamalan howler

Alouatta pigra

GENUS: Ateles: (spider monkey)…range: Tropical South and

Central America

SPECIES: Ateles belzebuth: longhaired spider monkey

Ateles fusciceps:brownhaired spider monkey

Ateles geoffroyi: black-handed spider monkey

Ateles paniscus: black spider monkey

GENUS: Brachyteles: (Wolley spider monkey)…range: Brazil

SPECIES: Brachyteles arachnoids: Wolley spider monkey

GENUS: Lagothrix: (Wolley monkeys)..range: Andes North of Peru

SPECIES: Lagothrix flavicauda : yellow-tailed wolley

Lagothrix lagothricha: common wolley

9. FAMILY: Cercopithecidae

a. Sub-Family: Cercopithecinae

GENUS: Macaca: (macaques)…range: SE Asia

SPECIES: Macaca arctoides: bear macaque

Macaca speciosa: stump-tailed macaque

Macaca assamensis : Assam macaque

Macaca fascicularis: crab-eating macaque

Macaca maurus: Moor macaque

Macaca mulatta: Rhesus macaque

Macaca nemestrina: pigtail macaque

Macaca nigra: Celebes ape

Macaca nigrescens : black ape

Macaca radiata: bonnet macaque

Macaca silenus: liontail macaque

Macaca sinica: torque macaque

Macaca sylvanus: barbary ape

Macaca thibetana: Tibetan stump-tailed

Macaque

Macaca tonkeana: Tonkean macaque

Macaca hecki

GENUS: Cercocebus: (mangabeys)…range: W. and E.

Africa

SPECIES: Cercocebus albigena: white-cheeked

mangabey

Cercocebus aterrimus: black mangabey

Cercocebus galeritus: Agile mangabey

Cercocebus agilis: Tana River mangabey

Cercocebus torquatas: white-collared mangabey

Cercocebus atys: sooty mangabey

GENUS: Papio: (baboons)…range: Sub-Saharan East Africa,

Arabian arid zones

SPECIES: Papio anubis: Olive baboon (sub-saharan E.

Africa)

Papio cynocephalus: yellow baboon

Papio hamadryas: Hamadryas baboon

Papio papio: Guinea baboon (Ethiopian savannas)

Papio ursinus: Chacma baboon (S. Africa)

GENUS: Mandrillus: (mandrill/drill)…range: forests of W.

Africa

SPECIES: Mandrillus leucophaeus: Drill

Mandrillus sphinx : Mandrill

GENUS: Theropithecus: (gelada)…range: Ethiopian

Highland Mtns.

(Gich and Sankaber areas of   Semien Mtns Nat'l Park)

SPECIES: Theropithecus gelada: Gelada

GENUS: Cercopithecus…range: East Africa, Mtns of W. Uganda,

W. C. Africa, Medium altitude forests

SPECIES: Cercopithecus aethiops: savanna monkey

Cercopithecus pygerythrus: green monkey

Cercopithecus tantalus: vervet

Cercopithecus sabaeus: grivet

Cercopithecus ascanius: Schmidt's guenon

Cercopithecus campbelli: Campbell's monkey

Cercopithecus cephus: Moustached monkey

Cercopithecus denti: Dents monkey

Cercopithecus wolfi

Cercopithecus diana: Diana monkey

Cercopithecus dryas:Dryas monkey

Cercopithecus erythrogaster: red-bellied monkey

Cercopithecus erythrotis: red-eared monkey

Cercopithecus hamlyni: owl-faced monkey

Cercopithecus hamlyni: L'Hoest's monkey

Cercopithecus mitis: Diademed monkey

Cercopithecus albogularis:Sykes' monkey

Cercopithecus mona: mona monkey

Cercopithecus neglectus: De Brazza's monkey

Cercopithecus nictitans: greater white-nosed Monkey

Cercopithecus petaurista: lesser white-nosed Monkey

Cercopithecus pogonias: crowned monkey

Cercopithecus preussi: Preuss' monkey

Cercopithecus lhoesti

Cercopithecus salongo: Zaire Diana monkey

Cercopithecus solatus: sun-tailed monkey

Cercopithecus wolfi: Wolf's monkey

GENUS: Miopithecus: (talapoi)...range: West Central Africa

SPECIES: Miopithecus talpoin: Talapoi

GENUS: Allenopithecus: (Allen's swamp monkey)...range: C.

Africa

SPECIES: Allenopithecus nigroviridis: Allen's swamp Monkey

GENUS: Erythrocebus: (patas monkey)...range: West, Central and

parts Of east Africa

SPECIES: Erythrocebus patas : patas monkey

b. Sub-Family: Colobinae

GENUS: Colobus: (colobus monkeys)...range: West , W. Central

and East Africa.   SPECIES: Colobus angolensis: Angolan colobus

Colobus badius: red colobus (pilocolobus)

Colobus rufomitratus: Bay colobus

Colobus guereza: Guereza (Eastern black and White colobus)

Colobus kirkii: Kirk's colobus

Colobus polykomos: King's colobus (Western

Black and white colobus)

Colobus satanas: black colobus

GENUS: Procolobus:(olive)…range: West Africa only

SPECIES: Procolobus verus: olive colobus

GENUS: Pygathrix:…range: all are limited to forests of Indochina

SPECIES: Pygathrix avunculus: Tokin snub-nosed monkey

Pygathrix brelichi: Brelichi's snub-nosed monkey

Pygathrix nigripes

Pygathrix roxellana: Chinese snub-nosed monkey

GENUS: Simias (pig-tailed langur)…range: limited to Mentawi

Islands west of Sumatra

SPECIES: Simias concolor: pig-tailed langur

GENUS: Nasalis: (proboscis monkey)…range: Coastal areas of Borneo

SPECIES: Nasalis larvatus: proboscis monkey

GENUS: Prebytis (leaf/langur monkeys)…range: S/SW India

and Sri Lanka.

SPECIES: Prebytis aurata

Prebytis comata : sundra leaf monkey

Prebytis hosei: grizzled leaf monkey

Prebytis thomasi: ebony leaf monkey

Prebytis cristata: silvered leaf monkey

Prebytis entellus: Hanuman langur

Prebytis francoisi: Francois monkey

Prebytis delacouri

Prebytis frontata: white-fronted leaf monkey

Prebytis geei: golden leaf monkey

Prebytis johnii: Nilgiri monkey

Prebytis melalophos: banded leaf monkey

Prebytis femoralis: mitred leaf monkey

Prebytis obscura: dusky leaf monkey

Prebytis phayrei: Phayre's leaf monkey

Prebytis pileata: capped leaf monkey

Prebytis potenziani: Mentawai leaf monkey

Prebytis rubicunda: maroon leaf monkey

Prebytis vetulus: purple-faced leaf monkey

Prebytis senex

10. FAMILY: Hylobatidae

GENUS: Hylobates: (gibbons and siamang)…range:

Rainforests of SE Asia   SPECIES: Hylobates

agilis: agile gibbon

Hylobates concolor: crested gibbon

Hylobates hoolock: hoolock gibbon

Hylobates klossii: Kloss' gibbon

Hylobates lar: common gibbon

Hylobates moloch: Javan gibbon

Hylobates muelleri: Muller's gibbon

Hylobates pileatus: Pileated gibbon

Hylobates syndactylus: Siamang

11. FAMILY: Pongidae

GENUS: Pongo: (orangutans)…range: Only Borneo and Sumatra

Islands

SPECIES: Pongo pygmaeus: orangutans

GENUS: Pan: (bonobos and chimpanzees)…range:

Bonobo: Congo River region

Chimps: Ethiopian: west and

Central equitorial Africa

SPECIES: Pan paniscus: bonobos

Pan troglodytes: chimpanzees

12. FAMILY: Hominidea

GENUS: Gorilla: (gorillas)...range:

Western lowland: Ethiopian tropical rainforests,

S. Nigeria to Congo

Mountain: Ethiopian Virunga Mtns separating Zaire From Rwanda
and Uganda.

GENUS: Homo

SPECIES: Sapiens: humans

Tree Shew Taxonomy

This list shows all of the extant tree-shrews

(Lyon,1913; Napier and Napier, 1967; cited in Martin, 1990).

Subfamily Ptilocercinae

Ptilocercus lowii

Pen-tailed tree-shrew

Subfamily Tupaiinae

Tupaia belangeri

Belanger's tree-shrew

Tupaia glis

Common tree-shrew

Tupaia longipes

Long-footed tree-shrew

Tupaia montana

Montane tree-shrew

Tupaia nicobarica

Nicobar tree-shrew

Tupaia picta

Painted tree-shrew

Tupaia palawanensis

Palawan tree-shrew

Tupaia splendidula

Rufous-tailed tree-shrew

Tupaia minor

Pygmy tree-shrew

Tupaia javanica

Indonesian tree-shrew

Tupaia gracilis

Slender tree-shrew

Anathana ellioti

Indian tree-shrew

Lyonogale tana

Terrestrial tree-shrew

Lyonogale dorsalis

Striped tree-shrew

Urogale everetti

Philippine tree-shrew

Dendrogale melanura

Southern smooth-tailed tree-shrew

Dendrogale murina

Northern smooth-tailed tree-shrew

This table shows supposed shared characteristics between tree-shrews and primates, from Martin (1990). Results of Analysis of Characters Context

Shared Similarities

Skull

1. Snout relatively short   1. Snout in fact secondarily elongated in tree-shrews

2. Simplified set of turbinal bones   2. Set of six turbinal bones probably

primitive for placental mammals

3. Enlarged, forward-facing orbits   3. Orbits relatively small and laterally facing

4. Postorbital bar present   4. Postorbital bar present as a convergent

development in various mammals

5. Pattern of bones in medial orbital wall   5. Palatine/lacrimal contact in medial orbital wall probably primitive

6. Well-developed jugal bone with foramen

6. Well-developed jugal bone with foramen probably primitive

7. Enlarged braincase   7. Braincase has become enlarged

convergently in various mammalian groups

8. Inflated auditory bulla containing 'free'   ectotympanic ring

8. Auditory bulla formed from   entotympanic, not from petrosal;

ectotympanic ring is primitively ring-shaped

in placental mammals

9. Internal carotid pattern (bony tubes) 9. Enclosure of internal carotid in bony

tubes probably primitive

10. 'Advanced' form of auditory ossicles

10. Auditory ossicles do not clearly share derived features with primates

Dentition

1. Tooth-comb present at front of lower

jaw, linked with a specialized, serrated

sublingua

1. Tooth-comb formed exclusively from

incisors as a convergent feature; sublingua

present in common ancestor of marsupials and placentals

2. Reduced dental formula

2. Convergent reduction of dental formula in many mammalian groups

3. Similarities in cheek teeth between tree-shrews and certain primates with relatively primitive cheek teeth (e.g. Tarsius)

3. Limited similarities in cheek teeth between tree-shrews and certain primates undoubtedly due to primitive retention

Postcranial Morphology

1. Limbs and digits highly mobile

1. Limbs and digits probably highly mobile in ancestral placental mammals

2. Numerous details of limb musculature

2. Limb musculature shares primitive retentions with prosimians

3 Osteological similarities in both forelimbs and hindlimbs

3 Osteological similarities in forelimbs and hindlimbs attributable to primitive retention from ancestral placental mammals

4. Ridged skin on palms and soles

4. Ridged skin on palms and soles possibly a primitive feature for placental mammals;

tree-shrews lack the characteristic Meissner's corpuscles of primates

Brain and Sense

Organs

1. Olfactory apparatus reduced

    1. Olfactory apparatus not reduced relative to body size in tree-shrews

2. Visual apparatus enhanced

    2. Visual apparatus mildly enhanced;numerous primate features lacking

3. Central, avascular area of retina

    3. Unusual, spoke-like radiation of retinal vessels; unusual innervation

4. Neocortex expanded; brain size increased

    4. Expansion of neocortex and brain size found in many mammals

5. Calcarine sulcus present

    5. Calcarine sulcus not present in the brain of tree-shrews

Reproductive Biology

1. Penis pendulous; testes scrotal

1. Pendulous penis and scrotal testes   present in many mammals

2. Discoidal placenta, as in tarsiers and simians

2. Discoidal placenta common in mammals; endotheliochorial in tree-shrews

3. Small litter size; small number of teats

3. Small litter size and small number of teats common in mammals; tree-shrew offspring   are altricial, not precocial like those of primates

Miscellaneous

1. Caecum present

1. Caecum probably a primitive feature of marsupials and placentals

2. Molecular affinities (e.g. albumins)

2. No convincing molecular affinities between tree-shrews and   primates

Table 3. This table shows some features that are shared amongst extant primates but are absent in tree-shrews. Adapted from Martin (1990).

Primate Condition   **Tree-shrew Condition**

hands and feet   prehensile   not prehensile

brain   lateral and calcarine sulcus present   lateral and calcarine sulcus absent

scrotum   postpenial   prepenial

lower incisors   two or less on each side of the lower jaw   three on each side of the

lower jaw

upper incisors

  arranged transversely *   arranged longitudinally

offspring at birth

  precocial   altricial

gestation period

relatively long compared to body size   relatively short compared to body size

many archaic primates have longitudinally arranged upper incisors (e.g. Plesiadapis

  tricuspidens)

Primate Ranges (Adapted From martin, 1990)

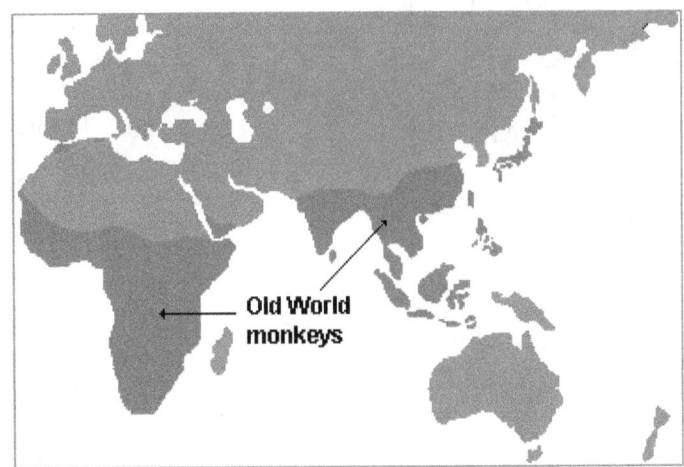

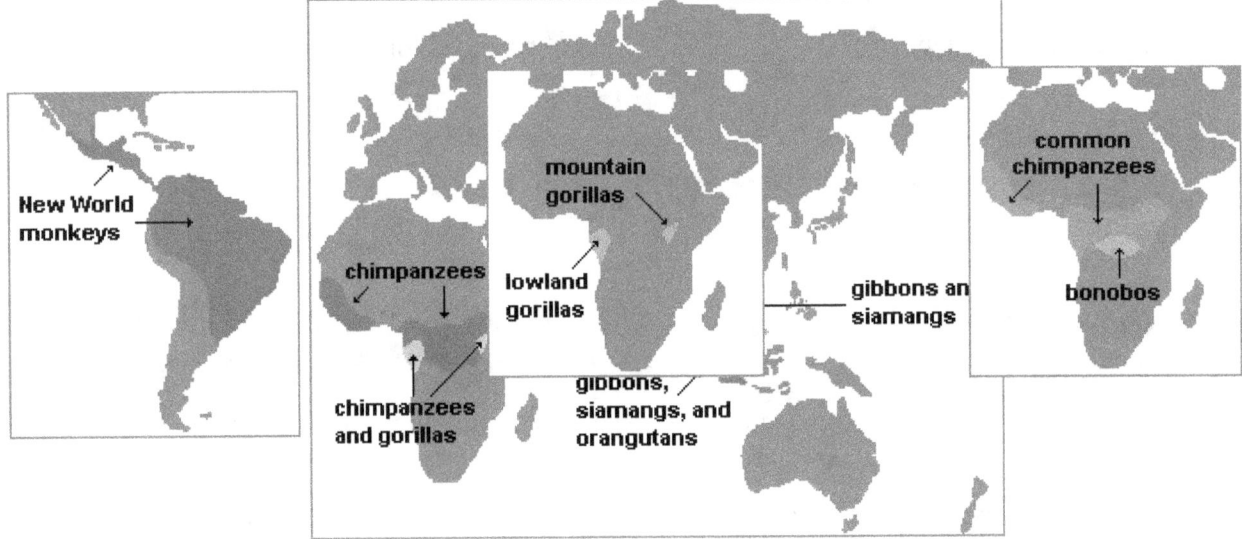

Geological Time

(From Aiello 1993)

The Geological Time Scale

Relative Time Span of Eras	Era	Period	Epoch	Age (Millions of Years Ago)	Some Important Events in the History of Life
CENOZOIC	CENOZOIC	Quaternary	Recent	0.01	Historic time
MESOZOIC			Pleistocene	1.8	Ice ages; humans appear
			Pliocene	5	Apelike ancestors of humans appear
PALEOZOIC		Tertiary	Miocene	23	Continued radiation of mammals and angiosperms
			Oligocene	35	Origins of many primate groups, including apes
			Eocene	57	Angiosperm dominance increases; origins of most modern mammalian orders
			Paleocene	65	Major radiation of mammals, birds, and pollinating insects
	MESOZOIC	Cretaceous		145	Flowering plants (angiosperms) appear; many groups of organisms, including most dinosaur lineages, become extinct at end of period
		Jurassic		208	Gymnosperms continue as dominant plants; dinosaurs dominant
		Triassic		245	Cone-bearing plants (gymnosperms) dominate landscape; first dinosaurs, mammals, and birds
	PALEOZOIC	Permian		290	Extinction of many marine and terrestrial organisms; radiation of reptiles; origins of mammal-like reptiles and most modern orders of insects
PRE-CAMBRIAN		Carboniferous		363	Extensive forests of vascular plants; first seed plants; origin of reptiles; amphibians dominant
		Devonian		409	Diversification of bony fishes; first amphibians and insects
		Silurian		439	Diversity of jawless fishes; first jawed fishes; colonization of land by vascular plants and arthropods
		Ordovician		510	Origin of plants; marine algae abundant
		Cambrian		570	Origin of most modern animal phyla
	PRECAMBRIAN			610	Diverse invertebrate animals; first vertebrates; diverse algae
				700	Oldest animal fossils
				1700	Oldest eukaryotic fossils
				2500	Oxygen begins accumulating in atmosphere
				3500	Oldest fossils known (prokaryotes)
				4600	Approximate time of origin of Earth

Age Based upon Dental Eruption and Developmental Information

(Park, 1999)

PERMANENT DENTITION

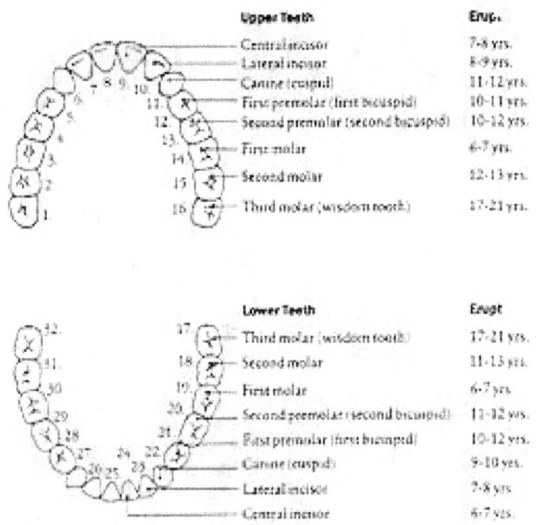

Upper Teeth	Erup.
Central incisor	7-8 yrs.
Lateral incisor	8-9 yrs.
Canine (cuspid)	11-12 yrs.
First premolar (first bicuspid)	10-11 yrs.
Second premolar (second bicuspid)	10-12 yrs.
First molar	6-7 yrs.
Second molar	12-13 yrs.
Third molar (wisdom tooth)	17-21 yrs.

Lower Teeth	Erup.
Third molar (wisdom tooth)	17-21 yrs.
Second molar	11-13 yrs.
First molar	6-7 yrs.
Second premolar (second bicuspid)	11-12 yrs.
First premolar (first bicuspid)	10-12 yrs.
Canine (cuspid)	9-10 yrs.
Lateral incisor	7-8 yrs.
Central incisor	6-7 yrs.

Classification of the Primates

This is the taxonomy Szalay and Delson (1979) proposed for the primates. The taxonomy only goes to the level of genus in the Linnean hierarchy.

PRIMATA

SUBORDER: Plesiadapiformes

SUPERFAMILY: Paromomyoidea

FAMILY: Paromomyidae

TRIBE: Purgatoriini

Purgatorius

TRIBE: Paromomyini

SUBTRIBE: Palaechthonina

Palaechthon

Plesiolestes

Palenochtha

SUBTRIBE: Paromomyina

Paromomys

Ignacius

Phenacolemur

TRIBE: Micromomyini

Micromomyini

Tinimomys

TRIBE: Navajoviini

Navajovius

Berruvius

FAMILY: Picrodontidae

Picrodus

Zanycteris

SUPERFAMILY: Plesiadapoidea

FAMILY: Plesiadapidae

Pronothodectes

Plesiadapis

Chiromyoides

Platychoerops

FAMILY: Saxonellidae

Saxonella

FAMILY: Carpolestidae

Elphidotarsius

Carpodaptes

SUBORDER: Strepsirhini

INFRAORDER: Adapiformes

FAMILY: Adapidae

SUBFAMILY: Notharctinae

Pelycodus

Notharctus

Smilodectes

Copelemur

SUBFAMILY: Adapinae

TRIBE: Protoadapini

Protoadapis

Agerinia

Europolemur

Mahgarita

Pronycticebus

TRIBE: Anchomomyini

Anchomomys

Huerzeleris

Periconodon

TRIBE: Microadapini

Microadapis

TRIBE: Adapini

SUBTRIBE: Adapina

Leptadapis

Adapis

SUBTRIBE: Caenopithecina

Caenopithecus

TRIBE: Indralorisini

Indraloris

FAMILY: incertae sedis

Amphipithecus

Lushius

INFRAORDER: Lemuriformes

SUPERFAMILY: Lemuroidea

FAMILY: Lemuridae

Lemur

Lepilemur

Hapalemur

Varecia

FAMILY: Megaladapidae

Megaladapis

SUPERFAMILY: Indrioidea

FAMILY: Indriidae

Indri

Propithecus

Avahi

Mesopropithecus

FAMILY: Daubentoniidae

Daubentonia

FAMILY: Archaeolemuridae

Archaeolemur

Hadropithecus

FAMILY: Palaeopropithecidae

Palaeopropithecus

Archaeoindris

SUPERFAMILY: Lorisoidea

FAMILY: Cheirogaleidae

Cheirogaleus

Phaner

Allocebus

Microcebus

FAMILY: Lorisidae

SUBFAMILY: Galaginae

Galago

Galagoides

Euoticus

Progalago

Komba

SUBFAMILY: Lorisinae

Loris

Nycticebus

Arctocebus

Perodicticus

Mioeuoticus

SUBORDER: Haplorhini

INFRAORDER: Tarsiiformes

FAMILY: Omomyidae

SUBFAMILY: Anaptomorphinae

TRIBE: Anaptomorphini

SUBTRIBE: Teilhardinina

Teilhardina

Chlorohysis

SUBTRIBE: Anaptomorphina

Anaptomorphus

SUBTRIBE: Tetoniina

Tetonius

Absarokius

Anemorhysis

Altanius

Mckennamorphus

TRIBE: Trogolemurini

Trogolemur

SUBFAMILY: Omomyinae

TRIBE: Omomyini

SUBTRIBE: Omomyina

Omomys

Chumashius

SUBTRIBE: Mytoniina

Ourayia

Macrotarsius

TRIBE: Washakiini

Loveina

Shoshonius

Washakius

Dyseolemur

Hemiacodon

TRIBE: Uintaniini

Uintanius

TRIBE: Utahiini

Utahia

Stockia

TRIBE: Rooneyiini

Rooneyia

SUBFAMILY: Ekgmowechashalinae

Ekgmowechashala

SUBFAMILY: Microchoerinae

Nannopithex

Necrolemur

Microchoerus

Pseudoloris

FAMILY: incertae sedis

Donrussellia

Hoanghonius

FAMILY: Tarsiidae

Tarsius

INFRAORDER: Platyrrhini

FAMILY: Cebidae

SUBFAMILY: Cebinae

Cebus

Saimiri

Neosaimiri

Dolichocebus

SUBFAMILY: Braniselllinae

Branisella

SUBFAMILY: Callitrichinae

TRIBE: Callitrichini

Callithrix

Cebuella

Saguinus

Leontopitheucs

TRIBE: Callimiconini

Callimico

FAMILY: Atelidae

SUBFAMILY: Atelinae

TRIBE: Atelini

Ateles

Lagothrix

Brachyteles

TRIBE: Alouattini

Alouatta

Stirtonia

SUBFAMILY: Pitheciinae

TRIBE: Pitheciini

SUBTRIBE: Callicebina

Callicebus

SUBTRIBE: Pitheciina

Pithecia

Chiropotes

Cacajao

Cebupithecia

SUBTRIBE: Xenotrichini

Xenothrix

SUBTRIBE: Homunculini

Aotus

Homunculus

Tremacebus

INFRAORDER: Catarrhini

SUPERFAMILY: Parapithecoidea

FAMILY: Parapithecidae

Parapithecus

Apidium

SUPERFAMILY: Cercopithecoidea

FAMILY: Cercopithecidae

SUBFAMILY: Cercopithecinae

TRIBE: Cercopithecini

Cercopithecus

Erythrocebus

Allenopithecus

TRIBE: Papionini

SUBTRIBE: Papionina

Papio

Cercocebus

Parapapio

Dinopithecus

Gorgopithecus

SUBTRIBE: Macacina

Macaca

Procynocephalus

Paradolichopithecus

SUBTRIBE: Theropithecina

Theropithecus

SUBFAMILY: Colobinae

SUBTRIBE: Colobina

Colobus

Libypithecus

Cercopithecoides

Paracolobus

Colobina, new genus

Colobina, gen. Indet.

SUBTRIBE: Semnopithecina

Presbyits

Pygathrix

Nasalis

SUBFAMILY: incertae sedis

Mesopithecus

Dolichopithecus

FAMILY: incertae sedis

Prohylobates

Victoriapithecus

FAMILY: Oreopithecidae

Oreopithecus

SUPERFAMILY: Hominoidea

FAMILY: Pliopithecidae

Propliopithecus

Pliopithecus

Dendropithecus

FAMILY: incertae sedis

FAMILY: Hominidae

SUBFAMILY: Hylobatinae

Hylobates

SUBFAMILY: Ponginae

TRIBE: Pongini

Pongo

Pan

TRIBE: Dryopithecini

Dryopithecus

TRIBE: Sugrivapithecini

Sivapithecus

Gigantopithecus

SUBFAMILY: Homininae

Ramapithecus

Australopithecus

Homo

Homo or Australopithecus sp(p). indet.

Homininae, gen. Et sp(p). indet.

INFRAORDER: incertae sedis

Pondaungia

Oligopithecus

REFERENCES:

Szalay, F. and Delson, E. 1979. Evolutionary History of the Primates.
Academic Press: New York.

This is the taxonomy Fleagle (1999) proposed for the primates. This is a

revision of the Fleagle (1988) taxonomy. The taxonomy only goes to the level
of genus in the Linnean hierarchy.

ORDER: incertae sedis

Purgatorius

ORDER: Plesiadapiformes

SUPERFAMILY: Plesiadapoidea

FAMILY: Plesiadapidae

Pandemonium

Pronothodectes

Nannodectes

Plesiadapis

Chiromyoides

Platychoerops

FAMILY: Carpolestidae

Chronolestes

Elphidotarsius

Carpodaptes

Carpolestes

Carpocristes

FAMILY: Saxonellidae

Saxonella

FAMILY: Paromomyidae

Paromomys

Ignacius

Phenacolemur

Elwynella

Simpsonlemur

Dillerlemur

Pulverflumen

Arcius

FAMILY: Micromomyidae

Micromomys

Tinimomys

Chalicomomys

Myrmekomomys

SUPERFAMILY: Microsyopoidea

FAMILY: Palaechthonidae

Palaechthon

Plesiolestes

Talpohenach

Torrejonia

Palenochtha

Premnoides

FAMILY: Microsyopidae

Navajovius

Berruvius

Niptomomys

Uintasorex

Avenius

Microsyops

Arctodontomys

Craseops

Megadelphus

SUPERFAMILY: incertae sedis

FAMILY: Picrodontidae

Picrodus

Zanycteris

Draconodus

FAMILY: Picromomyidae

Picromomys

Alveojunctus

ORDER: Primata

SUBORDER: Prosimii

FAMILY: incertae sedis

Altanius

Altiatlasius

INFRAORDER: Lemuriformes

SUPERFAMILY: Adapoidea

FAMILY: Notharctidae

SUBFAMILY: Notharctinae

Cantius

Copelemur

Notharctus

Smilodectes

Pelycodus

Hesperolemur

SUBFAMILY: Cercamoniinae

Donrussellia

Protoadapis

Europolemur

Periconodon

Caenopithecus

Pronycticebus

Cercamonius

Anchomomys

Huerzeleria

Buxella

Agerinia

Panobius

Mahgarita

Djebelemur

Aframonius

Omanodon

Shizarodon

Wadilemur

FAMILY: incertae sedis

Azibius

Hoanghonius

Lushius

Rencunius

Wailekia

FAMILY: Adapidae

SUBFAMILY: Adapinae

Adapis

Cryptadapis

Microadapis

Leptadapis

Adapoides

FAMILY: Sivaladapidae

SUBFAMILY: Sivaladapinae

Indraloris

Sivaladapis

Sinoadapis

SUPERFAMILY: Lemuroidea

FAMILY: Cheirogaleidae

Microcebus

Mirza

Cheirogaleus

Phaner

Allocebus

FAMILY: Lemuridae

Lemur

Eulemur

Varecia

Pachylemur

Hapalemur

FAMILY: Lepilemuridae

SUBFAMILY: Lepilemurinae

Lepilemur

SUBFAMILY: Megaladapinae

Megaladapis

FAMILY: Indriidae

SUBFAMILY: Indriinae

Avahi

Propithecus

Indri

SUBFAMILY: Archaeolemurinae

Archaeolemur

Hadropithecus

SUBFAMILY: Palaeopropithecinae

Mesopropithecus

Babakotia

Palaeopropithecus

Archaeoindris

FAMILY: Daubentoniidae

Daubentonia

SUPERFAMILY: Lorisoidea

FAMILY: Galagidae

Progalago

Komba

Otolemur

Galago

Euoticus

Galagoides

FAMILY: Lorisidae

Perodicticus

Pseudopotto

Arctocebus

Loris

Nycticebus

Mioeuoticus

FAMILY: Plesiopithecidae

Plesiopithecus

INFRAORDER: Tarsiiformes

SUPERFAMILY: Omomyoidea

FAMILY: Omomyoidae

SUBFAMILY: Anaptomorphinae

TRIBE: Anaptomorphini

Teilhardina

Anaptomorphus

Gazinius

Tetonius

Pseudotetonius

Absarokius

Tatmanius

Strigorhysis

Acrossia

TRIBE: Trogolemurini

Trogolemur

Sphacorhysis

Anemorhysis

Tetonoides

Arapahovius

Chlororhysis

TRIBE: Washakiini

Washakius

Shoshonius

Dyseolemur

Loveina

SUBFAMILY: Omomyinae

TRIBE: Omomyini

Omomys

Chumashius

Steinius

TRIBE: Uintaniini

Uintanius

Jemezius

TRIBE: Macrotarsiini

Macrotarsius

Hemiacodon

Yaquius

TRIBE: Ourayini

Ourayia

Wyomomys

Ageitodendron

Utahia

Stockia

Chipetaia

Asiomomys

TRIBE: incertae sedis

Ekgmowechashala

FAMILY: Microchoeridae

Nannopithex

Pseudoloris

Necrolemur

Microchoerus

FAMILY: incertae sedis

Rooneyia

Kohatius

SUPERFAMILY: Tarsioidea

FAMILY: Tarsiidae

Afrotarsius

Tarsius

Xanthorhysis

SUBORDER: Anthropoidea

INFRAORDER: incertae sedis

SUPERFAMILY: incertae sedis

FAMILY: Eosimiidae

Eosmias

FAMILY: incertae sedis

Amphipithecus

Pondaungia

Siamopithecus

SUPERFAMILY: Parapithecoidea

FAMILY: Parapithecidae

Serapia

Qatrania

Apidium

Parapithecus

Biretia

SUPERFAMILY: incertae sedis

FAMILY: incertae sedis

Proteopithecus

Arsinoea

Algeripithecus

Tabelia

INFRAORDER: Platyrrhini

SUPERFAMILY: Ceboidea

FAMILY: Atelidae

SUBFAMILY: Pitheciinae

Soriacebus

Carlocebus

Homunculus

Cebupithecia

Nuciruptor

Propithecia

Pithecia

Chiropotes

Cacajao

SUBFAMILY: Callicebinae

Callicebus

SUBFAMILY: Atelinae

Stirtonia

Protopithecus

Caipora

Alouatta

Lagothrix

Brachyteles

Ateles

FAMILY: Cebidae

SUBFAMILY: Aotinae

Tremacebus

Aotus

SUBFAMILY: Cebinae

Dolichocebus

Chilecebus

Neosaimiri

Laventiana

Cebus

Saimiri

SUBFAMILY: Callitrichinae

Micodon

Patasola

Lagonimico

Callimico

Saguinus

Leontopithecus

Callithrix

Cebuella

SUBFAMILY: incertae sedis

Branisella

Szalatavus

Mohanamico

Paralouatta

Xenothrix

Antillothrix

INFRAORDER: Catarrhini

SUPERFAMILY: Propliopithecoidea

FAMILY: Propliopithecidae

Propliopithecus

Aegyptopithecus

FAMILY: Oligopithecidae

Oligopithecus

Catopithecus

SUPERFAMILY: Hominoidea

FAMILY: Proconsulidae

Proconsul

Rangwapithecus

Limnopithecus

Dendropithecus

Simiolus

Micropithecus

Kalepithecus

Kamoyapithecus

Dionysopithecus

Platydontopithecus

FAMILY: Oreopithecidae

Mabokopithecus

Nyanzapithecus

Oreopithecus

FAMILY: incertae sedis

Afropithecus

Morontopithecus

Turkanapithecus

Kenyapithecus

Otavipithecus

Samburupithecus

FAMILY: Pliopithecidae

Pliopithecus

Plesiopliopithecus

Anapithecus

Laccopithecus

FAMILY: Hylobatidae

Hylobates

FAMILY: Pongidae

Dryopithecus

Lufengpithecus

Griphopithecus

Sivapithecus

Ankarapithecus

Gigantopithecus

Graecopithecus

Ouranopithecus

Pongo

Gorilla

Pan

FAMILY: Hominidae

SUBFAMILY: Australopithecinae

Ardipithecus

Australopithecus

Paranthropus

SUBFAMILY: Homininae

Homo

SUPERFAMILY: Cercopithecoidea

FAMILY: Victoriapithecidae

SUBFAMILY: Victoriapithecinae

Prohylobates

Victoriapithecus

FAMILY: Cercopithecidae

SUBFAMILY: Cercopithecinae

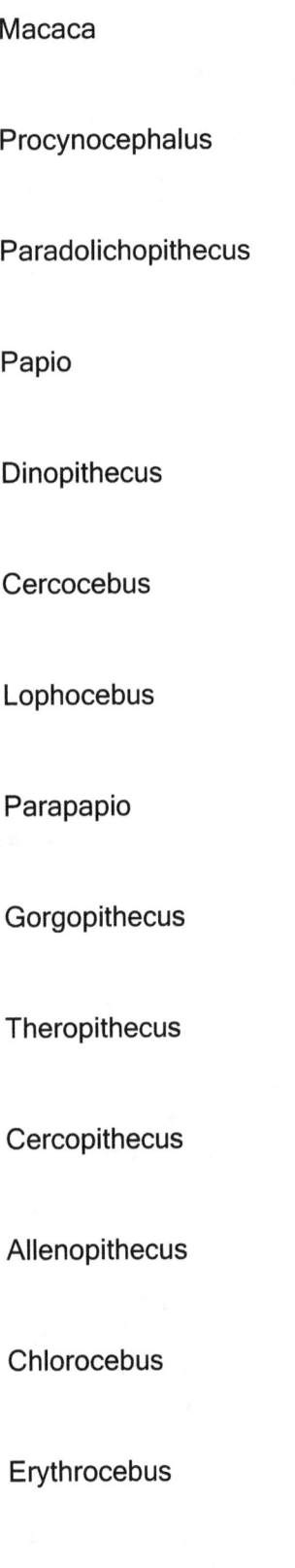

Macaca

Procynocephalus

Paradolichopithecus

Papio

Dinopithecus

Cercocebus

Lophocebus

Parapapio

Gorgopithecus

Theropithecus

Cercopithecus

Allenopithecus

Chlorocebus

Erythrocebus

Mandrillus

Miopithecus

SUBFAMILY: Colobinae

Mesopithecus

Dolichopithecus

Semnopithecus

Presbytis

Kasi

Trachypithecus

Nasalis

Simias

Pygathrix

Rhinopithecus

Colobus

Piliocolobus

Procolobus

Libypithecus

Microcolobus

Cercopithecoides

Paracolobus

Rhinocolobus

REFERENCES:

Fleagle, J.G. 1999. Primate Adaptation and Evolution. Academic Press: New York.

Hominid Phylogeny

(based upon Gibbons, 1981)

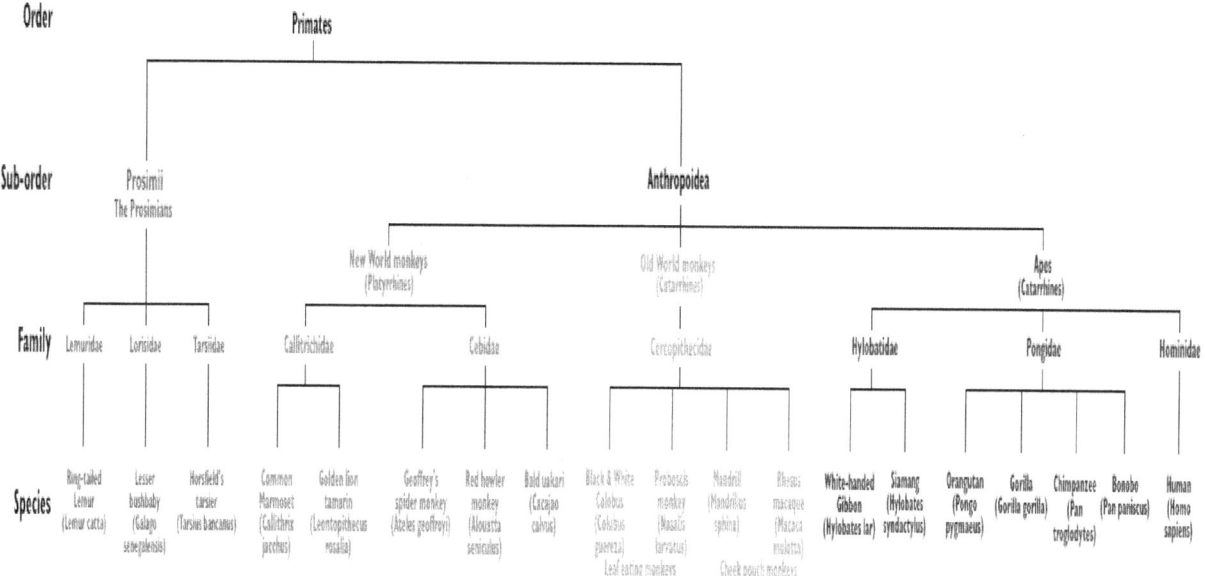

Australopithecus anamensis (4.4mya)

Australopithecus africanus (4.2mya)

                              Homo erectus (2.5mya)   *Australopithecus robustus*

(3.9mya)

                              Homo sapiens (1.5mya)

Homo sapiens neandertalensis (300,000ya)   *Homo sapiens sapiens* (150,000ya)

Primate Locomotor Anatomy

(Fleagle, 1999)

Arboreal Quadruped

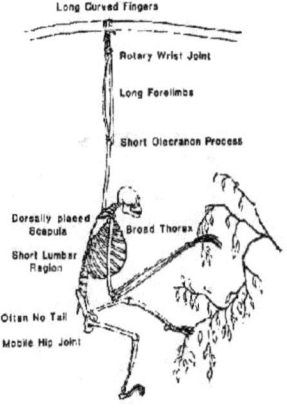

Long Tail

Narrow Thorax

Laterally-Placed Scapulae

Long Olecranon Process

Deep Ulna

Grasping foot

Short, similar length Forelimb and Hindlimb

Suspensory Primate

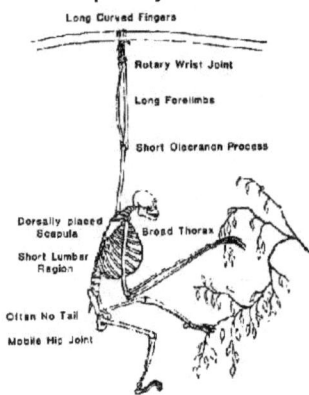

Long Curved Fingers

Rotary Wrist Joint

Long Forelimbs

Short Olecranon Process

Dorsally placed Scapula

Broad Thorax

Short Lumbar Region

Often No Tail

Mobile Hip Joint

Suspensory Primate

Long Curved Fingers

Rotary Wrist Joint

Long Forelimbs

Short Olecranon Process

Dorsally placed Scapula

Broad Thorax

Short Lumbar Region

Often No Tail

Mobile Hip Joint

Biped

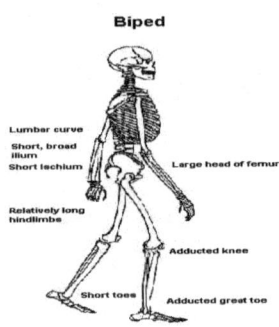

Lumbar curve

Short, broad ilium

Short ischium

Large head of femur

Relatively long hindlimbs

Adducted knee

Short toes

Adducted great toe

History of the Primates

(taken from original papers in the initial publications)

The Archaic Primates

Suborder Plesiadapiformes (Simons and Tattersall, 1972)

Superfamily Paromomyoidea (Simpson, 1940)

Family incertae sedis

Genus Purgatorius (Van Valen and Sloan, 1965) Cretaceous to Paleocene, N. America

Family Microsyopidae (Szalay, 1969a)

Genus Berruvius (Russel, 1964) Paleocene to Eocene, Europe

Genus Micromomys (Szalay, 1973) Paleocene to Eocene, N. America

Genus Navajovius (Matthew and Granger, 1921) Paleocene to Eocene, N. America

Genus Palaechthon (Gidley, 1923) Paleocene, N. America

Genus Palenochtha (Simpson, 1937) Paleocene, N. America

Genus Plesiolestes (Jepsen, 1930b) Paleocene, N. America

Genus Talpohenach (Kay and Cartmill, 1977) Paleocene, N. America

Genus Torrejonia (Gazin, 1968) Paleocene, N. America

Genus Alveojunctus ( ) Eocene, N. America

Genus Arctodontomys ( ) Eocene, N. America

Genus Craseops ( ) Eocene, North America

Genus Microsyops (Leidy, 1972) Eocene, N. America

Genus Niptomomys ( ) Eocene, N. America

Genus Tinimomys (Szalay, 1974b) Eocene, North America

Genus Uintasorex (Brown and Rose, 1976) Eocene, N. America

Family Paromoyidae (Simpson, 1940)

Genus Paromomys (Gidley, 1923) Paleocene, N. America

Genus Ignacius (Matthew and Granger, 1921) Paleocene to Eocene, N. America

Genus Phenacolemur (Matthew, 1915) Paleocene to Eocene, N. America, Europe

Genus Arcius () Eocene, Europe

Genus Elwynella () Eocene, N. America

Family Picrodontidae Simpson, 1937)

Genus Draconodus (Tjomida, 1982) Paleocene, N. America

Genus Picrodus (Douglass, 1908) Paleocene, . America

Genus Zanycteris (Matthew, 1917a) Paleocene, N. America

Superfamily Plesiadapoidea (Trouessart, 1897)

Family Plesiadapidae (Trouessart, 1897)

Genus Chiromyoides (Stehlin, 1916) Paleocene to Eocene, N. America, Europe

Genus Nannodectes Gingerich, 1974) Paleocene, N. America

Genus Platychoerops (Charlesworth, 1854) Eocene, Europe

Genus Plesiadapis (Gervais, 1877) Paleocene to Eocene, N. America, Europe

Genus Pronothodectes (Gidley, 1923) Paleocene, N. America

Family Saxonellidae (Russel, 1964)

Genus Saxonella (Russell, 1964) Paleocene, N. America, Europe

Family Carpolestidae (Simpson, 1935b)

Genus Elphidotarsius (Gidley, 1923) Paleocene, N. America

Genus Carpodaptes (Matthew and Granger, 1921) Paleocene, N. America

Genus Carpolestes (Simpson, 1928) Paleocene to Eocene, N. America

Eocene Primates

Order Primates (Linnaeus, 1758)

Suborder Prosimii (Illiger, 1811)

Infraorder Tarsiiformes (Gregory, 1915b)

Family Omomyidae (Trouessart, 1879)

Subfamily Anaptomorphinae (Cope, 1883)

Genus Absarokius (Matthew, 1915) Eocene, N. America

Genus Aycrossia () Eocene, N. America

Genus Altanius (Dashzeveg and McKenna, 1977) Eocene, N. America

Genus Anaptomorphus (Cope, 1872) Eocene, N. America

Genus Anemorhysis (Gazin, 1958) Eocene, N. America

Genus Chlororhysis (Gazin, 1958) Eocene, N. America

Genus Gazinius () Eocene, N. America

Genus Lovenia (Simpson, 1940) Eocene, N. America

Genus Mckennamorphus (Szalay, 1976) Eocene, N. America

Genus Steinius () Eocene, N. America

Genus Strigorhysis () Eocene, N. America

Genus Teilhardina (Simpson, 1940) Eocene, France

Genus Tetonius (Matthew, 1915) Eocene, N. America

Genus Trogolemus (Matthew, 1909) Eocene, N. America

Subfamily Omomyinae (Trouessart, 1879)

Genus Arapahovius () Eocene, N. America

Genus Chumashius (Stock, 1933) Eocene, N. America

Genus Dyseolemur (Stock, 1934) Eocene, N. America

Genus Hemiacodon (Marsh, 1872a) Eocene, N. America

Genus Jemezius () Eocene, N. America

Genus Macrotarsius (Clark, 1941) Eocene to Oligocene, N. America

Genus Omomys (Leidy, 1869) Eocene, N. America

Genus Ourayia (Gazin, 1958) Eocene, N. America

Genus Shoshonius (Granger, 1910) Eocene, N. America

Genus Stockia (Gazin, 1958) Eocene, N. America

Genus Uintanius (Matthew, 1915) Eocene, N. America

Genus Utahia (Gazin, 1958) Eocene, N. America

Genus Washakius (Leidy, 1873) Eocene, N. America

Subfamily Microchoerinae (Lydekker, 1887)

Genus Microchoerus (Wood, 1846) Eocene to Oligocene, England, France, Germany

Genus Nannopithex (Stehlin, 1916) Eocene, France, Germany

Genus Necrolemur (Fihol, 1873) Eocene, France

Genus Psleudoloris (Stehlin, 1916) Eocene, France, Spain

Family Omomyidae, incertae sedis

Genus Hoanghonius (Zdansky, 1930) Eocene, China

Infraorder Adapiformes (Szalay and Delson, 1979)

Family Adapidae (Trouessart, 1879)

Subfamily Notharctinae (Trouessart, 1879)

Genus Cantius (Simons, 1962b) Eocene, N. America

Genus Copelemur (Gingerich and Simons, 1977) Eocene, Rocky Mt.

Genus Notharctus (Leidy, 1870) Eocene, Rocky Mt.

Genus Pelycodus Cope, 1875) Eocene, N. Americas and Europe

Genus Smilodectes (Wortman, 1903) Eocene, Rocky Mt.

Subfamily Adapinae (Trouessart, 1879)

Genus Adapis (Cuvier, 1821) Eocene to Oligocene, Europe

Genus Agerinaia (Crusafont-Pairo and Golpe-Posse, 1973) Eocene, Spain

Genus Anchomomys (Stehlin, 1916) Eocene, Europe

Genus Caenopithecus (Rütimeyer, 1862) Eocene, Europe

Genus Cercamonius (Gingerich, 1975c) Eocene, Europe

Genus Cryptadapis () Eocene, Europe

Genus Donrussellia (Szalay, 1976) Eocene, Europe

Genus Europolemur (Weigelt, 1933) Eocene, German Democratic Republic

Genus Huerzeleris (Szalay, 1974a) Eocene, Europe

Genus Indraloris (Lewis, 1933) Miocene, Northern India

Genus Leptadapis (Gervais, 1876) Eocene, Europe

Genus Mahgarita (Wilson and Szalay, 1976) Eocene, Texas

Genus Microadapis (Szalay, 1974a) Eocene, Switzerland

Genus Periconodon (Stehlin, 1916) Eocene, Europe

Genus Pronycticebus (Grandidier, 1904) Eocene, France

Genus Protoadapis (Lemoine, 1878) Eocene, Western Europe

Family Adapidae, incertae sedis

Genus Lushius (Chow, 1961) Eocene, China

Suborder incertae sedis

Genus Amphipithecus (Colbert, 1937) Eocene, Burma

Oligocene Primates

Order Primates (Linnaeus, 1758)

Suborder Prosimii (Illiger, 1811)

Infraorder Tarsiiformes (Gregory, 1915b)

Family Omomyidae (Trouessart, 1879)

Subfamily Omomyinae (Trouessart, 1879)

Genus Rooneyia (Wilson, 1966) Oligocene, N. America

Subfamily Ekgmowechashalinae (Szalay, 1976)

Genus Edgmowechashala (Macdonald, 1963) Oligocene, N. America

Family Tarsiidae (Gray, 1825)

Genus Afrotarsius (Simons and Brown, 1985) Oligocene, Africa

Suborder Anthropoidea (Mivart, 1864)

Infraorder Platyrrhini (E. Geoffroy, 1812)

Superfamily Ceboidea (Simpson, 1931)

Family Cebidae (Bonaparte, 1831)

Subfamily Cebinae (Bonaparte, 1831)

Genus Dolichocebus (Kraglievich, 1951) Oligocene, Argentina

Subfamily Branisellinae (Hershkovitz, 1977)

Genus Branisella (Hoffstetter, 1969) Oligocene, Bolivia

Family Atelidae (Gray, 1825)

Subfamily Pitheciinae (Mivart, 1865)

Genus Tremacebus (Hershkovitz, 1974a) Oligocene, Argentina

Infraorder Catarrhini E. Geoffroy, 1812)

Superfamily Parapithecoidea (Schlosser, 1911)

Family Parapithecidae (Schlosser, 1911)

Genus Parapithecus (Schlosser, 1910) Oligocene, Egypt

Genus Apidium (Osborn, 1908) Oligocene, Egypt

Genus Qatrania (Simons and Kay, 1983) Oligocene, Egypt

Genus Catopithecus (), Oligocene, Egypt

Superfamily Hominoidea (Gray, 1825)

Family Propliopithecidae (Straus, 1961)

Subfamily Propliopithecinae (Delson and Andrews, 1975)

Genus Propliopithecus (Schlosser, 1916) Oligocene, Egypt

Genus Aegyptopithecus (Simons, 1965) Oligocene, Egypt

Subfamily Oligopithecinae

Genus Catopithecus (Simons, 1989) Oligocene, Egypt

Genus Proteopithecus (Simons, 1989) Oligocene, Egypt

Genus Oligopithecus (Simons, 1962a) Oligocene, Egypt

Miocene Primates

Order Primates (Linnaeus, 1758)

Suborder Prosimii (Illiger, 1811)

Infraorder Adapiformes (Szalay and Delson, 1979)

Family Adapidae (Trousessart, 1879)

Subfamily Sivaladapinae

Genus Indraloris (Lewis, 1933) Miocene, Asia

Genus Sivaladapis () Miocene, Asia

Genus Sinoadapis () Miocene, Asia

Infraorder Lemuriformes (Gregory, 1915b)

Superfamily Lorisoidea (Gray, 1821)

Family Lorisidae (Gray, 1821)

Subfamily Galaginae (Gray, 1825)

Genus Progalago (Macinnes, 1943) Miocene, East Africa

Genus Komba (Simpson, 1967) Miocene, East Africa

Subfamily Lorisinae (Flower and Lydekker, 1891)

Genus Mioeuoticus (Leakey, 1962) Miocene, East Africa

Genus Nycticeboides () Miocene, Asia

Infraorder Tarsiiformes (Gregory, 1915b)

Family Tarsiidae (Gray, 1825)

Genus Tarsius (Storr, 1789) Miocene to Modern, Asia

Family Omomyidae (Trouessart, 1879)

Subfamily Anaptomorphinae (Cope, 1883)

Genus Ekgmowechashala (Macdonald, 1963)

Miocene, N. America

Suborder Anthropoidea (Mivart, 1864)

Infraorder Platyrrhini (E. Geoffroy, 1812)

Superfamily Ceboidea (Simpson, 1931)

Family Cebidae (Bonaparte, 1831)

Subfamily Cebinae (Bonaparte, 1831)

Genus Neosaimiri (Stirton, 1951) Miocene, Colombia

Subfamily Callitrichinae

Genus Micodon () Miocene, Colombia

Subfamily Aotinae ()

Genus Tremacebus (Hershkovitz, 1974a) Miocene, Argentina

Genus Aotus (Illiger, 1811) Miocene, Colombia

Genus Homunculus (Ameghino, 1891) Miocene, Argentina

Family Atelidae (Gray, 1825)

Subfamily Atelinae (Gray, 1825)

Genus Stirtonia (Hershkovitz, 1970a) Miocene, Colombia

Subfamily Pitheciinae (Mivart, 1865) Miocene, South America

Genus Cebupithecia (Stirton and Savage, 1951) Miocene, Colombia

Genus Mohanamico () Miocene, Colombia

Infraorder Catarrhini E. Geoffroy, 1812)

Superfamily Cercopithecoidea (Gray, 1821)

Family Cercopithecidae (Gray, 1821

Subfamily Cercopithecinae (Gray, 1821)

Genus Parapapio (Jones, 1937) Miocene to Pleistocene, Africa

Genus Macaca (Lacépéde, 1799) Miocene to modern, Europe, N. Africa, Asia

Subfamily Colobinae (Blyth, 1875)

Genus Colobus (Illiger, 1811) Miocene to modern. Africa

Genus Libypithecus (Stromer, 1913) Miocene, Egypt

Genus Microcolobus () Miocene, East Africa

Genus Presbytis (Eschscholtz, 1821) Miocene to modern, Asia

Genus Mesopithecus (Wagner, 1839) Miocene to Pliocene, Europe

Family Victoriapithecidae

Genus Prohylobates (Fourtau, 1918) Miocene, Egypt, Libya

Genus Victoriapithecus (von Koenigswald, 1969) Miocene, Kenya, Uganda

Superfamily Hominoidea (Gray, 1825)

Family Oreopithecidae (Schwalbe, 1915)

Genus Nyanzapithecus (Harrison, 1987) Miocene, East Africa

Genus Oreopithecus (Gervais, 1872) Miocene, Italy

Family Pliopithecidae (Zapfe, 1961a)

Genus Pliopithecus (Gervais, 1849) Miocene, Europe

Genus Laccopithecus (Wu and Pan, 1984) Miocene, China

Genus Dendropithecus (Andrews and Simons, 1977) Miocene, Kenya

Genus Crouzelia (Ginsburg, 1975) Miocene, Europe

Family Proconsulidae ()

Genus Proconsul (Hopwood, 1933a) Miocene, East Africa

Genus Limnopithecus (Hopwood, 1933a) Miocene, Africa

Genus Dendropithecus (Andrews, Pilbeam and Simons, 1976) Miocene, Africa

Genus Simiolus (R.E.F. Leakey and Leakey, 1987) Miocene, Africa

Genus Rangwapithecus (Pilbeam et al., 1977) Miocene, Africa

Genus Micropithecus (Fleagle & Simons, 1978) Miocene, Africa

Genus Dionysopithecus () Miocene, Asia

Genus Platydontopithecus () Miocene, Asia

Family Hylobatidae (Gray, 1870)

Genus Hylobates (Illiger, 1811)

Pleistocene to modern, South East Asia, Southern China

Family Pongidae (Elliot, 1913)

Genus Dryopithecus (Lartet, 1856) Miocene, Africa and Europe

Genus Lufengpithecus () Miocene, Asia

Genus Ouranopithecus () Miocene, Europe

Genus Sivapithecus (Pilgrim, 1910) Miocene, Africa, Europe, Asia

Genus Ramapithecus (Lewis, 1934) Miocene, Africa, Europe, Asia

Genus Gigantopithecus (von Koenigswald, 1935) Miocene to Pleistocene, Asia

Genus Graecopithecus (von Koenigswald, 1972) Miocene, Europe

Family incertae sedis

Genus Turkanapithecus (Leakey and Leakey, 1986a) Miocene, East Africa

Genus Afropithecus (Leakey and Leakey, 1986b) Miocene, East Africa Genus

Kenyapithecus (L. Leakey, 1962a)

Miocene, East Africa

Pliocene Primates

Order Primates (Linnaeus, 1758)

Suborder Prosimii (Illiger, 1811)

Suborder Anthropoidea (Mivart, 1864)

Infraorder Catarrhini E. Geoffroy, 1812)

Superfamily Cercopithecoidea (Gray, 1821)

Family Cercopithecidae (Gray, 1821

Subfamily Cercopithecinae (Gray, 1821)

Genus Cercopithecus (Linnaeus, 1758) Pliocene to modern, Africa

Genus Papio (Müller, 1773) Pliocene to modern, Africa

Genus Cercocebus (E. Geoffroy, 1812) Pliocene to modern, Africa

Genus Dinopithecus (Broom, 1937) Pliocene, Africa

Genus Gorgopithecus (Broom and Robinson, 1949a) Pliocene, South Africa

Genus Procynocephalus (Schlosser, 1924) Pliocene, Central China, N. India

Genus Paradolichopithecus (Necrasov, Samson, and Radules, 1976) Pliocene, Europe

Genus Theropithecus (I. Geoffroy, 1843)Pliocene to modern, Africa

Subfamily Colobinae (Blyth, 1875)

Genus Cercopithecoides (Mollett, 1947) Pliocene to Pleistocene, Africa

Genus Paracolobus (R. Leakey, 1969)Pliocene, East Africa

Genus Rhinocolobus() Pliocene to Pleistocene, Africa

Genus Dolichopithecus (Deperet, 1889) Pliocene to Miocene, Europe

Superfamily Hominoidea (Gray, 1825)

Family Hominidae (Gray, 1825)

Genus Australopithecus (Dart, 1925) Pliocene to Pleistocene, Africa

Genus Homo (Linnaeus, 1758) Pliocene to modern, Worldwide

Note: Species that continue into the Pleistocene are not repeated in the Pleistocene table.

Pleistocene and subfossil Primates

Order Primates (Linnaeus, 1758)

Suborder Prosimii (Illiger, 1811)

Infraorder Lemuriformes (Gregory, 1915b)

Superfamily Lemuroidea (Gill, 1872)

Family Lemuridae (Gray, 1821)

Genus Varecia (Gray, 1863)

Living and subfossil , Madagascar

Family Megaladapidae (Flower and Lydekker, 1891)

Genus Megaladapis (Major, 1894)

Subfossil , Madagascar

Superfamily Indriodea (Burnett, 1828)

Family Indriidae (Burnett, 1828)

Genus Mesopropithecus (Standing, 1905)

Subfossil , Madagascar

Family Daubentoniidae (Gray, 1863)

Genus Daubentonia (E. Geoffroy, 1795)

Modern and subfossil , Madagascar

Family Archaeolemuridae (Major, 1896)

Genus Archaeolemur (Filhol, 1895)

Subfossil , Madagascar

Genus Hadropithecus (Lorenz, 1899)

Subfossil , Madagascar

Family Palaeopropithecidae (Tattersall, 1973)

Genus Palaeopropithecus (G. Grandidler, 1899)

Subfossil , Madagascar

Genus Archaeoindris (Standing, 1908)

Subfossil , Madagascar

Suborder Anthropoidea (Mivart, 1864)

Infraorder Platyrrhini (E. Geoffroy, 1812)

Superfamily Ceboidea (Simpson, 1931)

Family Atelidae (Gray, 1825)

Subfamily Pitheciinae (Mivart, 1865)

Genus Xenothrix (Williams and Koopman, 1952)

Subfossil, Jamaica

`Infraorder Catarrhini E. Geoffroy, 1812)

Superfamily Cercopithecoidea (Gray, 1821)

Family Cercopithecidae (Gray, 1821

Subfamily Cercopithecinae (Gray, 1821)

Subfamily Colobinae (Blyth, 1875)

Genus Rhinopithecus

Pleistocene to modern, Asia

Superfamily Hominoidea (Gray, 1825)

Family Hylobatidae (Gray, 1870)

Genus Hylobates (Illiger, 1811)

Pleistocene to modern, South East Asia, Southern China

Family Pongidae (Elliot, 1913)

Genus Pongo (Lacepede, 1799)

Pleistocene to modern, South China to Borneo and Sumatra

PLIOCENE AND EARLY PLEISTOCENE HUMANS

Ardipithecus ramidus. White, Suwa & Asfaw, 1994

Date: 4.4 MYBP   Distribution: Africa   Sites: Aramis

Synonyms: Australopithecus ramidus

Australopithecus (Dart, 1925)

Australopithecus anamensis

Date: 3.9 to 4.2 MYBP

Distribution: Africa (Figure 13-5 map)

Sites: Kanapoi

Australopithecus afarensis (Johanson, White, and Coppens, 1978)

Date: 5 to 3 MYBP

Distribution: Africa

Australopithecus africanus (Dart, 1925)

Date: 3 to 2 MYBP

Distribution: Africa

Australopithecus robustus (Broom, 1938)

Date: 2 to 1.5 MYBP

Distribution: Africa

Sites: Kromdraai, Swartkrans (SK 46, SK 48,)

Synonyms: Paranthropus robustus (Broom, 1938); Paranthropus crassidens (Broom, 1949)

Australopithecus boisei (L. Leakey, Tobias, and Napier, 1964)

Date: 2.8 to 1.4 MYBP

Distribution: Africa

Sites: Olduvai Gorge (OH 5), Peninj, Lake Turkana (KNM-ER 406, KNM-ER 732,

KNM-WT

17,000)

Synonyms: Paraustralopithecus aethiopicus (Arambourg and Coppens), 1968; Homo

aethiopicus (Olson, 1985)

Homo (Linnaeus, 1758)

Homo habilis (L. Leakey, Tobias, and Napier, 1964)

Date: 2.4 to 1.5 MYBP

PLEISTOCENE HUMANS

Homo erectus (Dubois, 1892)

Date: 1.8 to 0.2 MYBP

synonyms: Pithecanthropus erectus (Dubois 1893); Hylobates giganteus (Bumuller,

1899); Sinanthropus pekinensis (Black and Zdansky, 1927); Homo modjokertensis (von

Koenigswald, 1936); Atlanthropus mauritanicus (Arambourg, 1954); Homo leakeyi

(Heberer, 1963); Tchadanthropus uxoris (Coppens, 1965); Homo ergaster (Groves and

Mazak, 1975)

Homo sapiens (Linnaeus, 1758)

Homo sapiens soloensis (Oppenoorth, 1932)

Date: ca 400,000 to 100,000 YBP

Distribution: Africa, Asia, Europe

omo sapiens rhodesiensis (Woodward, 1921)

Date: ca 0.4 to 0.1 MYBP

Distribution: Africa, Europe

Sites: Bodo, Elandsfontien; Kabwe; Petralona; ?Singa

Synonyms: Cyphanthropus rhodesiensis (Pycraft, 1928), Homo kanamensis (L. Leakey, 1935), Palaeoanthropus njarensis (Reck and Kohl-Larsen, 1936), Africanthropus njarensis (Weinert, 1938)

omo sapiens neanderthalensis (King, 1864)

Date: 135,000 to 29,000 YBP

Distribution: Circum-Mediterranean

Sites: Amud, Combe-Grenal, Devil's Tower, Ehringsdorf, Forbes' Quarry, Jebel Irhoud, Krapina, La Chapelle-aux-Saints, La Ferrassie, Le Moustier, La Quina, Monte Circeo, Neanderthal, Saccopastore, Shanidar, Tabun, Teshik-Tash

Synonyms: Protanthropus atavus (Haeckel, 1895); Homo europaeus primigenius (Wisler, 1898); Paleanthropus krapiniensis (Sergi, 1911); Homo primigenius (Schwalbe, 1903); Homo antiquus (Adloff, 1908);Homo transprimigenius mousteriensis (Forrer, 1908),

Homo neanderthalensis (Bonarelli, 1909);Homo priscus (Krause, 1909); Palaeanthropus

europaeus (Sergi, 1910); Homo calpicus (Keith, 1911);Archanthropus (Arldt, 1915);

Anthropus neanderthalensis (Boyd-Dawkins, 1926);Metanthropus (Sollas, 1933);

Pithecanthropus neanderthalensis (Sklerj, 1937)

Homo sapiens sapiens (Linnaeus, 1758)

Date: ca. 100,000 YBP to Holocene

Distribution: Africa, Asia, Europe

Sites: ?Border Cave, Combe-Capelle, Cro-Magnon, Fish Hoek, Grimaldi Jebel Qafzeh

[#6], Mugharet es Skhul [Skhul 5], Omo (Omo 1), Predmosti; Zhoukoutian Upper Cave,

Synonyms: Homo sapiens fissilis (Gorjanovic-Kramberger, 1905); Homo grimaldii

(Lapouge, 1906); Notanthropus eurafricanus recens (Sergi, 1911); Notanthropus

eurafricanus archaius (Sergi, 1911); Homo mediterraneus fossilis (Behm, 1915); Homo

capensis (Broom, 1917); Homo sapiens cromagnonensis (Gregory, 1921); Homo larterti

(Pycraft, 1935); Palaeanthropus palestinus (McCown and Keith, 1932 in Weidenreich

1932b).

Bibliography

Introduction

Archer, J.C., and Cracraft, J., 1992. "Phylogenetic Concepts". Genetics. 134:1299-1303.

Anderson, R.C., 1996. "Historical Development of the Concept of Adaptation." In: Rose, M.R., and Lauder, G.V., (eds). Adaptation. Academy Press. San Diego, CA.

Anderson, R.C. 1998. Toward a Sustainable Enterprise: The Interface Model. Chelsea Green Publishing, VT.

Bearder, S.K., 1997. "Bushbabies and Tarsiers:Diverse Societies in Solitary Foragers."In: Primate Societies. Smuts, B.B., Cheney, D.L., Seyfart, R.M., Wrangham, R.W., and Struhsaker, T.T., (eds). University of Chicago Press.

Berggren, W.A., Kent, C.C., Swisher, M., and Aubry, P., 1995. A Revised Conozoic Geochronology and Chronostratigraphy. SEPM (Society for Sedimentary Geology) Special Publication. 54:129-212.

Birdsell, J.B., 1972. Human Evolution. Rand McNally and Company, Chicago, IL.

Bown, T.M., 1976. "Affinites of Teilhardina (Primates, Omomyidae) with Description of a New Species from North America."Folia Primatologica. 25:62-72.

Bown, T.M., and Rose, K.D., 1984. Folia Primatologica 43:97-112.

Bown, T.M., and Rose, K.D., 1991. Journal of Human Evolution. 20:465-480.

Conroy, G.C., 1990. Primate Evolution. Norton. London, NY.

Conroy, G.C., 1992. Primate Evolution. Norton. London, NY.

Eldredge, N., 1980. "Newly Described Taxa in Micronesica." <u>Homola Dickinsoni</u> 16(2) 74-277.

Eldredge, N., 1989. <u>Microevolutionary Dynamics: Species, Niches, and Adaptive Peaks.</u> McGraw Hill. NY.

Fleagle, J.G., Bown, T.M., Obradovich, J.D., and Simons, E.L., 1986. "How Old are the Fayum Primates?" In: Else, J.G., and Lee, P.C. (eds). Primate Evolution. Cambridge University Press. NY.

Fleagle, J.G., 1988. <u>Primate Adaptation and Evolution.</u> Academic Press. San Diego, CA.

Fleagle, J.G., 1999. <u>Primate Adaptation and Evolution.</u> Academic Press. San Diego, CA.

Gibbons, M.F., Jr., 1972. "What is a Primate?: A Study in Current Taxonomic Thinking." <u>Discovery.</u> 8:1

Gibbons, M.F., Jr., 1981. "Marsupials –Their Place in the Natural History of Mammals." <u>Biology Digest.</u> 8:2.

Gibbons, M.F., Jr., 1999. "Biological Classification / Taxonomy / Systematics / Biosystematics : A Penultimate Draft. (non published)."

Gradstein, F.M., Agherberg, F.P., Ogg, J.G., Hardenbol, J., Van Veen, P., Thierry, J., and Huang Zehui, 1995. "A Triassic and Cretaceous Time Scale." In: Berggren, W.A., Kent, D.V., Aubrey, M.P., and Hardenbol, J., (eds). Geochronology, Time Scales and Global Stratigraphic Correlation. SEPM (Society for Sedimentary Geology) Special Publication. 54:95-126.

Griffiths, M., 1978. <u>The Biology of Monotremes.</u> Academic Press. NY.

Groves, C.P., 1977. <u>A Theory of Human Evolution and Primate Evolution.</u> Clarendon. Oxford.

Groves, C.P., 1989. <u>The Biology of Race.</u> University of California Press. Berkely.

Hershkovitz, P., 1977. <u>Living New World Monkeys (Platyrrhini) with and Introduction to the Primates. Volume one.</u> University of Chicago Press, IL.

Hill, W.G., 1972. "Effective Size of Populations with Overlapping Generations." <u>Theoretical Population Biology.</u> 3:279-289.

Kimball, W.J., 1993."Functional Adaptations." <u>Journal of Physiology.</u> 741-765.

Mirvart, G.J., 1873. <u>Man and Apes.</u> London.

Obradovich, J.D., 1993. "A Cretaceous Time Scale." In: <u>Evolution of the Western Basin.</u> Caldwell, W.G.E., (ed). Geological Association of Canada. Special Paper. 39:379-396.

Park, M.A., 1999. <u>Biological Anthropology, 2nd edition.</u> Mayfield. CA.

Pasqual, F., and Murphy, R., 1992. "Prehistoric Change". <u>Journal of Human Evolution.</u> 4:335-355.

Quirk, J.A. , and Donohue, M., 1983. <u>Evolutionary Advantages.</u> Columbia University Press.

Rosen, S.I., 1974. <u>Introduction to the Primates:Living and Fossil.</u> Prentice Hall. NJ.

Rowe, N., 1996. <u>The Pictorial Guide to the Living Primates.</u> Pognonias Press.NY.

Simons, E., 1972. <u>Primate Evolution.</u> MacMillan. NY.

Simpson, G.G., 1945. <u>The Principles of Classification and a Classification of Mammals.</u> Bulletin of the American Museum of Natural History. 85(ixvi).

Simpson, G.G., 1949. <u>The Meaning of Evolution: A Study of the History of Life and of its Significance for man.</u> Yale University Press. CT.

Simpson, G.G., 1953. The Main Features of Evolution. Columbia University

Press. NY.

Szalay, F. and Delson, E., 1979. <u>Evolutionary History of the Primates.</u> Academic

Press. NY.

Szalay, F., Rosenberger, and Dagosto, M., 1987. "Diagnosis and Differentiation   of

the Order Primates." <u>Yearbook of Physical Anthropology.</u> 30:75-105.

Vaughan, T.A., 1986. <u>Mammology, 3rd edition.</u> Saunders College Publishing. PA.

Walker, A., and Teaford, M., 1989. "The Hunt for Proconsul." <u>Scientific</u> <u>American.</u> 261:76-82.

Models of the Evolutionary Process

Charlesworth, B., 1990. "Mutation-Selection Balance and the Evolutionary Advantage of Sex and Recombination." <u>Genetic Research.</u> 55:199-221.

Eldredge, N., and Gould, S.J., 1972. "Punctuated Equilibria: An Alternative to Phyletic Gradualism." In: <u>Models in Paleobiology.</u> Schoff, T.J.M. , (ed). Freeman Cooper.San Francisco, CA.

Hoffman, A., 1989. <u>Arguments on Evolution: A Paleontologist's Perspective.</u>

Oxford University Press.

Moller, A.P. and Pomiankowski, A., 1993. "Why Have Birds Got Multiple Sexual

Ornaments?" <u>Behavioral Ecological Sociobiology.</u> 32:167-176.

Rice, C.M., Fuchs, R., Higgins, D.G., Stoehr, P.J., and Cameron, G.N., 1993.   "The

EMBL Data Library." <u>Nucleic Acid Research.</u> 21:2967-2971.

Simpson, G.G., 1953. <u>The Main Features of Evolution.</u> Columbia University

Press. NY.

Weiner, J., 1995. <u>The Beak of the Finch.</u> Knopf Publishing.

Arborealism

Begun, D.R., Ward, C.V., and Rose, M.D., (eds), 1997. <u>Function, Phylogeny, and Fossils:</u>

Miocene Hominoid Evolution and Adaptations. Plenum Press. NY.

Biegert, J., 1963. "The Evaluation of Characteristics of the Skull, Hands, and Feet for

Primate Taxonomy." In: Classification and Human Evolution. Washburn, S.L., (ed). Aldine

Press. Chicago, IL.

Cartmill, M., 1974. "Rethinking Primate Origins." Science. 184:436-443.

Conroy, G.C., 1990. Primate Evolution. Norton. NY.

Fleagle, J.G., 1988. Primate Adaptation and Evolution. Academic Press. NY.

Gregory, W.K., 1951. Evolution Emerging: A Survey of Changing patterns From

Primeval Life to Man. MacMillan. NY.

Hennig, W., 1965. "Phylogenetic Systematics." Annual Review of Entomology.

10: 97-116.

Park, M.A., 1999. <u>Biological Anthropology, 2nd edition</u>. Mayfield. CA.

Continental Drift

Bridges, T., 1990. "Rare Earth Tagging Methods for the Study of the Larval Dispersal by Marine Invertebrates." <u>Limology and Oceonography.</u> 38: 346-360.

Burrett, C.F., 1974. "Plate Tectonics and the Hercynian Orogeny." <u>Nature.</u> 239:155-157.

Churkin, M., Jr., 1973. "Paleozoic and Precacamitrian Rocks of Alaska and Their Role in the Structural Evolution of the U.S." <u>Ecological Survey Paper.</u> 7(10):6.

Creer, K.M., 1970. "A Review of Paleomagnetism." Earth Science Reviews. 6.
DeBlij, 1996a. <u>Physical Geography of the Global Environment, 2nd edition</u>. Wilcy. NY.

DuToit, A.L., 1937. <u>Our Wandering Continents: A Hypothesis of Continental Drift.</u> Hafner Publishing. NY.

Fleagle, J.G., 1988. <u>Primate Adaptation and Evolution.</u> Academic Press. San Diego, CA.

Fleagle, J.G., 1999. <u>Primate Adaptation and Evolution.</u> Academic Press. San Diego, CA.

Hume, W.F., 1867. "Terrestrial Theorie: A Digest of Various Views as to the Origin and Development of the Earth and Their Bearing on the Geology of Egypt." Cairo Gov't Press.

Irving, E., and Irving, G.A., 1977. "Apparent polar Wander Paths:Carbiniferous through Cenozoic and the Assembly of Gondwana." Geophysical Surveys. 5:141-188.

Laing, D., 1991. The Earth System: An Introduction to Earth Science. WC Brown. Iowa.

Morel, H.J.,Wilcox, B.P., and Wood, M.K., 1948. "Factors Influencing Infiltration and Erosion in the Gaudalupe Mountains of New Mexico." In: Proceedings AGU Front Range Branch Hydrology Days. Morel-Seytoux, H.J., and Warner, J.W., (eds). 6(11):122.

Raymo, C.,1983. The Crust of the Earth. Prentice Hall. NJ.

Tarling, D. , and Tarling, M., 1975. Continental Drift: A Study of the Earth's Moving Surface. Anchor Books.

Windley, 1984. The Evolving Continents. Wiley, NY.

Rooneyia

Bown, T.M., 1976. "Affinites of Teilhardina (Primates, Omomyidae) with Description of a New Species from North America."<u>Folia Primatologica</u>. 25:62-72.

Fleagle, J.G., 1988. <u>Primate Adaptation and Evolution</u>. Academic Press. NY.

Gunnell, G.F., 1995. "New Notharctine (Primates, Adapiformes) Skull from the Uintan (Middle Eocene) of San Diego County, California." <u>American Journal of Physical Anthropology.</u> 98:447-470.

Rose, K.D., and Bown, T.M., 1991. "Additional Fossil Information on the Differentiation Of the Oldest Euprimates." Proceedings of the National Academy of Science. 88:98-101.

Wilson, 1996. <u>Measuring and Monitoring Diversity:Mammals.</u>

Primate Origins

Beard, K.C., 1990. "Gliding Behavior and Paleoecology of the Alleged Primate Family Paromomyidae (Mamalis, Dermoptera)." <u>Nature.</u> 345:340-341.

Bown, T.M., 1976. "Affinities of Teilhardina (Primates, Omomyidae) with Description of a New Species from North America."Folia Primatologica. 25:62-72.

Bown, T.M., and Rose, K.D., 1987. "Patterns of Dental Evolution in Early Euocene Anaptomorphine Primates (Omomyidae) From the Bighorn Basin, Wyoming." Paleontological Society. 23:1-162.

Buckely, G.A., 1997. "A New Species of Purgatorius (Mammalia ;Primatophora) From the Lower Paleocene Bear Formation. Crazy Mountain Basis, South Central Montana." Journal of Paleontology. 71:149-155.

Cartmill, M., 1972. "Arboreal Adaptations and the Origin of the Order Primates." In: The Functional and Evolutionary Biology of Primates. Tuttle, R., (ed). Aldine. IL.

Cartmill, M., 1974. "Rethinking Primate Origins". Science.184:436-443.

Conroy, G.C., 1990. Primate Evolution. WW Norton. NY.

Dolhinow, P., 1972. Primate Patterns, 1st edition. Holt, Rhinehart and Winston. NY.

Fleagle, J.G., 1988. Primate Adaptation and Evolution. Academic Press. San Diego, CA.

Fleagle, J.G., 1999. Primate Adaptation and Evolution. Academic Press. San Diego, CA

Gingerich, P.D., 1976. "Cranial Anatomy and Evolution of Early Tertiary Plesiadapidae (Mammalisa, Primates)." University of Michigan, Museum of Paleontology, Papers on Paleontology no. 15.

Gingerich, P.D., 1986. "Plesiadapids and the Deliniation of the Order Primates."In:Major Topics in Primate and Human Evolution. Wood, B., Martinand, L., and Andrews, P., (eds). Cambridge University Press.

LeGros Clark, W., 1934. Early Forunners of Man. London.

MacPhee, R.D.E., 1993. Primates and Their Relatives in Phylogenetic Perspectives(Advances in Primatology). Plenium Press.NY.

Rose, K.D., 1995. "Anterior Dentition and Relationships of the Early Euocen Omomyids.Arapahovias Advena and Teilhadina Demissa." Journal of Human Evolution. 28:231- 244.

Sussman, R.W., and raven, P.H., 1978. "Pollination by Lemurs and Marsupials. An Archaic

Coevolutionary System." <u>Science.</u> 200:731-736.

Szalay, F.S., 1972. "Paleobiology of the Earliest Primates." In: <u>The Functional and</u>

<u>Evolutionary biology of Primates.</u> Tuttle, R., (ed). Aldine. IL.

Van Valen, J.R., and Sloan, R.E., 1965. "The Earliest Primates." <u>Science.</u> 150:743-745.

Wible, J.R., and Covert, H.H., 1987. "Primates: Cladistic Diagnosis and

Relationship." <u>Journal of Human Evolution.</u> 16:1-20.

Tree Shrews

Carlsson, A., 1992. "Uber Die tupaiidae und I HRE Bexiehungen zu Den I insectivora und

Den Prosimiae. (translated)." <u>Acta Zoology.</u> Stockholm 3:227-270.

Fleagle, J.G., 1988. <u>Primate Adaptation and Evolution</u>. Academic Press. NY.

Gibbons, M.F., Jr., 1981. "The Taxonomic Significance of the Ovarian Bursa in the Tree Shrew (Mammalia : Tupaiidae)." Endicott Research Paper no. 1. Hanley, W, and Stratton, J. (eds).

Gibbons, M.F., Jr., 1999. "Biological Classification / Taxonomy / Systematics / Biosystematics: A Penultimate Draft." (non published).

Hill, W.C.O., 1953. "The Female Reproductive Organs or Tarsius with Observations on the Physiological Changes Herein." Proceedings of the Zoological Society of London. 123:589-598.

Huxley, T.H., 1872. The Manual of the Anatomy of Vertebrated Animals. Appleton.NY.

LeGros Clark, W., 1926. "On The Anatomy of the Pentailed Shrew *Ptilocercus lowii*." Proceedings of the Zoolgical Society of London. 46:1179-11809.

LeGros Clark, W., 1934. Early Forunners of Man. London.

LeGros Clark, W., 1960 (1959). The Antecedents of Man. Edinburgh University Press.

Edinburgh.

Schultz, A.H., 1969. <u>The Life of Primates</u>. Universe Books. NY.

Simpson, G.G., 1945. "The Principles of Classification and a Classification of

Mammals." <u>Bulletin of the American Museum of Natural History. 85(ixvi).</u>

Wbblew, J.R., and Zeller, U., 1994. "Cranial Circulation of the Pentailed Tree Shrew

Ptilocercus lowii and Realtionships of Scandentia." Journal of Mammalian Evolution.

2(4):209-230.

<u>Early Primates</u>

Andrews, P, and Martin, M., 1987. "Cladistic Relationships of Extant Fossil Hominoids."

<u>Journal of Human Evolution. </u>16:101-116.

Beard, K.C., Krishtalka, L., and Stucky, R.K., 1991. "First Skulls of the early Eocene

Primates *Shoshonius cooperi* and the Anthropoid Tarsier Dichotomy." Nature. 349: 64-66.

Delson, E., 1981. "Paleoanthropology:Pliocene and Pleistocene Human Evolution."

Paleobiology. 7:298-305.

Fleagle, J.G., 1988. Primate Adaptation and Evolution. Academic Press. NY.

Groves, C.P., 1989. The Biology of Race. University of California Press. Berkeley.

Herskovitz, P., 1977. Living New World Monkeys (Platyrrhini) Volume one. University

    of Chicago Press.

McHenry, H., 1975. "Fossils and the Mosaic Nature of Human Evolution." Science.

190:425-431.

Park, M.A., 1999. Biological Anthropology, 2nd edition. Mayfield. CA.

LeGros Clark, W., 1934. <u>Early Forunners of Man</u>. London.

LeGros Clark., W., 1960 (1959). <u>The Antecedents of Man</u>. Edinburgh University Press.

Edinburgh.

Parapathecidae

Bown, T.M., and Harrell, J.A., 1995. "The Oldest Paved Road, Fayum Depression,

Egypt." <u>The Ostracon</u>. 6(3):1-4.

Harrison, T., 1997. "The Phylogenetic Relationships of Early Catarrhine Primates: A

Review of the Current Evidence." <u>Journal of Human Evolution.</u> 16:41-80.

Howell, F.C., 1977. "Hominidae." In: Evolution of African Mammals. Magio, V.,

And Cooke, H.B.S., (eds). Cambridge and Harvard University Press.

Kay, R., and Simons, E., 1980. "The Ecology of Oligocene African Anthropoidea."

International Journal of Primatology. 1:21-37.

Olson, T., 1981. "Basicranial Morphology of the Extant Hominoids and Pliocene

Hominids: The New Material from the Hadar Formation, Ethiopia, and its

Significance in Early Human Evolution and Taxonomy." In Aspects of Human

Evolution. Stringer, C, (ed). Taylor and Francis. London.

Park, M.A., 1999. Biological Anthropology, 2nd edition. Mayfield. CA.

Tuttle, R., 1988. "What's New in African Paleoanthropology?" Annual Review of

Anthropology. 17:391-426.

Catarrhines

Andrews, P., 1985. "Hybrid Frequencies". Physical review Letters, no 54, 2022.

Fleagle, J.G., and Jungers, W., 1982. Fifty Years of Higher Primate Phylogeny. A History of American Physical Anthropology, 1930-1980. Spencer. NY.

Fleagle, J.G., 1988. Primate Adaptation and Evolution. Academic Press. San Diego, CA.

Fleagle, J.G., 1999. Primate Adaptation and Evolution. Academic Press. San Diego, CA.

Kay, R.F., Plavcan, J.M., Glander, K.E., and Wright, P.C., 1981. "Behavioral and Size Correlates of Canine Dimorphism in Platyrrhine Primates." American Journal of Physical Anthropology. 77(3):385-397.

Schwartz, J., Tattersall, I, and Eldredge, N., 1978. "Phylogeny and Classification of the Primates Revisited." Yearbook of Physical Anthropology. 21:95-133.

Simons, E., 1995. "Fossil Skulls , Jaws Shed New Light on Origins of Anthropoids." Science. June 20, 1995.

Szalay, F., and Delson, E., 1979. Evolutionary History of the Primates. Academic Press. NY.

Platyrrhines

Ciochon, R.L and Chiarelli, A.B., 1980. Evolutionary Biology of the New World Monkeys and Continental Drift. Perseus, Cambridge, MA.

Fleagle, J.G., 1999. Primate Adaptation and Evolution. Academic Press, San Diego, CA.

Houle, A., 1999. "The Origin of Platyrrhines: An Evolution of the Arctic Scenario and the Floating Island Model." American Journal of Physical Anthropology. 109(4):541-549.

Rosenberger, A,L., Setoguchi, T., and Hartwig, W.C., 1991. "Laventiana annectens, new genus and species: fossil evidence for the origin of callitrichine New World monkeys." Proceedings of the National Academy of Sciences. March 15;88(6):2137-2140.

Rosenberger, A.L., 1992. "Evolution of Feeding Niches in New World Monkeys." American Journal of Physical Anthropology. 88:525-562.

Simons, E., 1976. "The Fossil Record of Primate Phylogeny." In: Molecular Anthropology. Goodman, M, Tashian, R.E. and Tashian, J.H. editors. Plentium Press, New York

Simpson, G.G., 1980. Splendid Isolation. Yale University Press, New Haven , CT.

Miocene Radiation of the Apes

Howell, F.C., 1977. "Hominidae". In: Evolution of African Mammals. Magio, V.,

And Cooke, H.B.S., (eds). Cambridge and Harvard University Press.

McHenry, H., 1975. "Fossils and the Mosaic Nature of Human Evolution." Science.

190:425-431.

Pilbeam, D., 1988. Human Origins and Evolution. Cambridge University Press.

Rosen, S.I., 1974. Introduction to the Primates, Living and Fossil. Prentice Hall. NJ.

Simons, E., 1989. "Human Origins." Science. 245:1343-1350.

Skelton, R., McHenry, H., and Drawhorn, G. 1986. "Phylogenetic Analysis of Early

Hominids." Current Anthropology. 27:21-43.

Miocene Paleoecology

Alvarez, W., 1997. T. rex and the Crater of Doom. Princeton University Press.

Alvarez, H., Aparici, J., May, J., and Olmos, F., 1997. "Emplacement of Cretaceous-

Tertiary Boundry: Shocked Quartz from Chicxulub Crater." Science. 269:930-935.

Aubrey, M.P., Berggren, D.V., kent, J.J., Flynn, K.D., Klitgord, J.D., Obradovich, J.D., and

Prothero, D.R., 1988. "Paleogene Geochronology: An Integrated Approach."

Paleooceanography. 3:707-742.

Dodd, J.R., and Stanton, R.J., 1990. Paleoecology, Concepts and Applications, 2nd

edition. Wiley-Intersciences. NY.

Foster, J.H., and Opdyke, N.D., 1970. "Upper Miocene to Recent Magnetic Stratigraphy in Deep Sea Sediments." Journal of Geophysical Research. 75:4465-4473.

Harland, W.B., Armstrong, R.L., Cox, A.V., Craig, L.E., Smith, A.G., and Smith, D.G., 1990. A Geological Time Scale, 1989 edition. Cambridge University Press.

Hayes, D.E., 1972. Antarctic Oceanology II. The Australian-New Zealand Sector. American Geophysical Union 1972, Antarctic Research Series no. 19.

Park, M.A., 1999. Biological Anthropology, 2nd edition. Mayfield. CA.

Staski, E., and Marks, J., 1992. Evolutionary Anthropology: An Introduction to PhysicalAnthropology. Harcourt Brace. TX.

Lothagam Mandible

Hill, A., Ward, S., and Brown, B., 1992. "Anatomy and Age of the Lothagam Mandible." Journal of Human Evolution. 22:439-451.

Leakey, M.G., Feibel, C.S., Bernor, R.L., Harris, J.M., Cerling, T.E., Stewart, K.M., Storrs, G.W., Walker, A., Werdelin, L., and Winkler, A.J., 1996. "Lothagam: A Record of Faunal Change in the Late Miocene of East Africa." Journal of

Vertebrate Paleontology. 16(3): 556-570.

McDougall, I., and Fiebel, C.S., 1999. "Numerical Age Control for the Miocene-
Pliocene Succession at Lothagam, a Hominoid-Bearing Sequence in the
Northern Kenya Shift." Journal of the Geological Society. July. 156(4):731.

Stewart, K.M., 1997. "A New Species of Sindacharax (Teleostei:Characidae) From
Lothagam, Kenya, and Some Implications for the Genus." Journal of Vertebrate
Paleontology. 17(1):34-38.

White,T.D., 1980. "Additional Fossil Hominids From Laetoli, Tanzania:1976-1979
Specimens." American Journal of Physical Anthropology. 53:487-504.

Available HTTP: **http://www.calle.com/world/uganda/Lo.html**   (Lothagam altitude
and coordinate information site).

Proconsul

Fleagle,J.G., 1988. Primate Adaptation and Evolution. Academic Press. New York, N.Y.

Schwartz, J., Tattersall, I., and Eldredge, N., 1978. "Phylogeny and Classification of the

Primates Revisited." Yearbook of Physical Anthropology. 21:95-133.

Scott, J., and Symons, N., 1974. Introduction to Dental Anatomy. Edinburgh. Churchill
Livingston.

Simmons, E., 1972. Primate Evolution. Macmillan. New York, N.Y.

Szalay, F., and Delson, E., 1979. Evolutionary History of the Primates. Academic Press.

New York, N.Y.

Sivapithecus

Benefit, B.R., and McCrossin, M.L., 1995. "Miocene Hominoids and Hominid Origins." Annual Review of Anthropology. Annual 1995. 24:237-257.

Beynon, A.D., Dean, M.C., and Reid, D.J., 1991. "On Thick and Thin Enamel in Hominoids (Primate tooth formation:A Symposium)." American Journal of Physical Anthropology. 86(2):295-310.

Grine, F.E., and Ungar, P.S., 1991. "Incisor Size and Wear in *Australopithecus africanus* and *Paranthropus robustus*." Journal of Human Evolution. 20:313-340.

Martin, L., 1985. "Significance of Enamel Thickness in Hominoid Evolution." Nature. 314:260-263.

Pilbeam, D. R., Rose, M.D., Barry, J.C., and Ibrahim Shah, S.M., 1990. "New Sivapithecus Humeri from Pakistan and the Relationship of Sivapithecus and Pongo (comparing humerous bones from a fossil ape and the orangutan)." Nature. 348(6298):237-239.

Scott, R.S., Kappelman, J., and Kelley, J., 1999. "The Paleoenvironment of Sivapithecus parvada." Journal of Human Evolution. 36:245-274.

Simons, E.L., and Pilbeam, D.R., 1965. "Some Problems of Hominid Classification. (a broad overview from the mid 1960's. Origin and diversification of hominids)." American Scientist. 53:237-246.

Dryopithecus

Andrews, P., and Pilbeam, D., 1996. "The Nature of the Evidence." <u>Nature.</u> 379:123-124.

Anemone, R.L., and Watts, E.S., 1992. "Dental Development in Apes and Humans: a comment on Simpson, Lovejoy and Meindl (1990)." <u>Journal of Human Evolution.</u> 22:149-153.

Beynon, A.D., and Dean, M.C., 1988. "Distinct Dental Development Patterns in Early Fossil Hominids." <u>Nature.</u> 335:509-514.

Grine, F., 1987. "On the Eruption Pattern of the Permanent Incisors and First Permanent Molars in Paranthropus." <u>American Journal of Physical Anthropology</u>. 72:353-359.

Holloway, R.L., 1967. "Tools and Teeth: Some Speculations Regarding Canine Reduction." <u>American Anthropologist</u>. 69:63-67.

Pilbeam, D., 1978. "Rearranging Our Family Tree." <u>Human Nature</u>. 1(6):38-45.

Pilbeam, D., 1986. "Hominoid Evolution and Hominoid Origins." <u>American Anthropologist.</u> 88:295-312.

Simmons, E. L., 1972. <u>Primate Evolution</u>. Macmillan. New York, N.Y.

Szalay, F., and Delson, E., 1979. <u>Evolutionary History of the Primates</u>. Academic Press. New York, N.Y.

Ramapithecus

Andrews, P., and Cronin, J., (1982). "The Relationships of Sivapithecus and Ramapithecus and the Evolution of the Orangutan." <u>Nature.</u> 297:541-546.

Lewin, R., (1984). "The Dethroned Ape: Divergence of Man and Ape Rules Out Ape as Ancestor of Man." Science. 226 :1182-1183.

Lewin, R. (1988). <u>Bones of Contention</u>. New York, Simon and Schuster.

Pilbeam, D., (1966). "Notes on Ramapithecus, The Earliest Known Hominid, and Dryopithecus" <u>American Journal of Physical Anthropology</u>. July, 25(1):1-5.

Pilbeam,D., (1969). "Newly Recognized Mandible of Ramapithecus." <u>Nature.</u> June, 222(198):1093.

Pilbeam, D., (1978). "Rearranging Our Family Tree." <u>Human Nature</u>. June.
Sarich, V.M., and Wilson, A.C., (1967). "Immunological Time Scale For Hominid Evolution." <u>Science.</u> 128:1200-1203.

Staski, E., and Marks, J., (1992). <u>Evolutionary Anthropology: An Introduction to Physical Anthropology</u> and Archaeology. Harcourt Brace.TX.

Thorne, A., and Wolpoff M.H., (1992). "The Multiregional Evolution of Humans." <u>Scientific American.</u>266 (4): 76-82.

Wu, R.K., and Oxnard, C.E., (1983). "Ramapithecus From China: Evidence From Tooth Dimensions." <u>Nature</u>. 306(5940):258-260.

Agyptopithecus

Day, M.H., 1970. <u>All Color Guide to Fossil Man</u>. Grosset and Dunlap. NY.

Olson, T., 1981. "Basicranial Morphology of the Extant Hominoids and Pliocene Hominids: The New Material from the Hadar Formation, Ethiopia, and its Significance in Early Human Evolution and Taxonomy." <u>In Aspects of Human Evolution</u>. Stringer, C, (ed). Taylor and Francis. London.

Simons, E., 1970. "The Origin and Radiation of the Primates" Annals of New York Academy of Sciences. 167. 319.

Simons, E., 1972. <u>Primate Evolution</u>. MacMillan. NY.

Gigantopithecus

Birdsell, J.B., 1972. <u>Human Evolution:An Introduction to the New Physical Anthropology.</u> Rand McNally. Chicago, IL.

Day, M.H., 1970. <u>All Color Guide to Fossil Man.</u> Grossett and Dunlap, NY.

LeGros Clark, W., 1955 (1954). The Fossil Evidence for Human Evolution. University of Chicago Press.

LeGros Clark, W., 1960 (1959). <u>The Antecedents of Man.</u> Edinburgh University Press. Ediburgh.

LeGros Clark, W., 1965. <u>History of the Primates.</u> Chicago University Press.

Simons, E., 1972. <u>Primate Evolution: An Introduction to Man's Place in Nature.</u> MacMillan. NY.

Simpson, G.G., 1949. <u>The Meaning of Evolution: A Study of the History of Life and of its Significance for Man.</u> Yale University Press. CT.

Von Koenigswald,G.H.R., 1952. Gigantopithecus blacki, a Giant Fossil Hominoid from the Pleistocene of Southern China." Anthropological Paper of the American Museum of Natural History. 43: 293-325.

Oreopithecus

LeGros Clark, W., 1960 (1959). <u>The Antecedents of Man.</u> Edinburgh University Press

Ediburgh.

Delson, E., 1979. "Oreopithecus is a Cercopithecoid After All." <u>American Journal of Physical Anthropology</u> 50(3):431-432.

Delson, E., 1981. "Paleoanthropology:Pliocene and Pleistocene Human Evolution." <u>Paleobiology</u>. 7:298-305.

Fleagle, J.G., 1999. <u>Primate Adaptation and Evolution.</u> Academic Press. CA.

Harrison, T., 1986. "New Fossil Anthropoids from the Middle Miocene of East Africa and Their Bearing on the Origin of the Oreopithecidae." <u>American Journal of Physical Anthropology.</u> 71:265-284.

Hutchinson, G.E., 1965.<u>The Evolutionary Theatre and the Evolutionary Play.</u> Yale University Press.

Hurzeler, J. 1954. <u>Zur Systemtischen Stellung Von OreopithecusVerhandlungen Der</u>

<u>Natureforschenden Gesellschaft in Basel.</u> Naturforschenden Gesellschaft in Basel, Basel

(translated).

Pilbeam, D., 1972. The Ascent of Man: An Introduction to Human Evolution. The

Macmillan Series in Physical Anthropology. New York: Macmillan Publishing Co., Inc.

Simons, E., 1964. "On the Mandible of Ramapithecus." <u>Proceedings of the</u>

<u>NationalAcademy of Sciences.</u> 51:528-535.

Simons, E., 1970. "The Origin and Radiation of the Primates" Annals of New York Academy of Sciences. 167. 319.

Simons, E., 1972. <u>Primate Evolution: An Introduction to Man's Place in Nature.</u> MacMillan. CA.

Afropithecus

Fleale, J.G., 1999. <u>Primate Adaptation and Evolution</u>. Academic Press.CA.

Leakey, R.E., and Leakey, M.G., 1986. "A New Miocene Hominoid from Kenya."

<u>Nature</u>. 324:143-146.

Leakey, M.G., and Walker, A., 1997. "Afropithecus: Function and Phylogeny." In:

<u>Function, Phylogeny and Fossils: Miocene Hominoid Evolution and Adaptation</u>.Begun,

D.R., Ward, C.V., and Rose, M.D., (eds). Plenum.NY.

Rose, M.D., 1993. "Functional Anatomy of the Primate Elbow and Forearm." In:

<u>Postcranial Adaptation in Nonhuman Primates.</u> Gebo, D.(ed). Northern Illinois

University Press.

Simons, E. 1989. "Human Origins." <u>Science</u>. 245:1343-1350.

<u>Morotopithecus</u>

Andrews, P., 1992. "Evolution and Environment in the Hominoidea." <u>Nature</u>. 360:641-

646.

Andrews, P., and Pilbeam, D., 1996. "The Nature of the Evidence.". <u>Nature.</u> 379:123-4.

Bishop, W., 1963. "The Late Tertiary and Pleistocene in Eastern Equitorial Africa." In:

<u>African Ecology and Human Evolution.</u> Howell, F., and bourliere, F., (eds). Aldine. IL.

Bishop, W., and Allbrook, D.,1963. "New Fossil Hominoid Material From Uganda."

<u>Nature</u>. 197:1187-1190.

Delson, E., 2000. "Morotopithecus." <u>Encylopedia of Human Evolution and Prehistory, 2nd</u>

<u>edition.</u> Delson, E., Tattersall, I., Van Couvering, J.A., and Brooks, A.S., (eds). Garland,

NY.

Gebo, D.G., MacLatchy, L., Kityo, R., Deino, A., Kingston, J., and Pilbeam, D., 1997. "A

New Hominoid Genus from the Uganda Early Miocene." <u>Science.</u> 276. 401-404.

LeGros Clark, W., and Leakey, L., 1951. <u>Fossil Mammals of Africa No. 1.</u> "The

Miocene Hominoidea of East Africa." British Museaum of Natural History. London.

Maclatchy, L., Gebo, D.G., Kityo, R., and Pilbeam, D., 2000. "Postcranial Functional

Morphology of Morotopithecus bishopi, with Implications for the Evolution of Modern Ape

Locomotion." Journal of Human Evolution.39(2):159-83.

Pickford M., Senut B. and Gommery D., 1999. "Sexual dimorphism in *Morotopithecus*

bishopi, an early Middle Miocene hominoid from Uganda and a reassessment of its

geological and biological contexts." In : <u>Late Cenozoic Environments and Hominid Evolution</u>

<u>: a tribute to Bill Bishop.</u> Andrews, P., and Banham, P., (eds). Geological Society,

Londres.

Pilbeam, D., and Simons, E. , "Some Problems of Hominid Classification," <u>American</u>

<u>Scientist</u> 53:237.

Pilbeam, D., 1969. "Tertiary Pongidae of East Africa: Evolutionary Relationships and

Taxonomy." <u>Bulliten 31</u>. Peabody Museum of Natural History. Yale Univeristy. CT.

Simons, E. , Pilbeam, D., 1965. "A Preliminary Revision of the Dryopithecinae

(Pongidae, Anthropoidea)." Folia Primatologica. 3:81-152.

Early Bipedalism

Aiello, L., and Dean, C., 1990. An Introduction to Human Evolutionary Anatomy.

Academic Press. CA.

Andrews, P, and Martin, L., 1987. "Cladistic Relationships of Extant Primates and Fossil

Hominoids." Journal of Human Evolution. 16:101-118.

Bates, M., 1964. Man in Nature, 2nd edition. Prentice Hall. NJ.

Binford, L., 1968. "Post-Pleistocene Adaptations." In: New Perspectives in Archeology.

Binford, S.R., and Binford, L.R., (eds). Aldine. IL.

Childe, V.G.,1939. Man Makes Himself. Oxford University Press.

Gibbons, M.F., Jr., 1981. "The Human Ancestry." Biology Digest.

Gibbons, M.F., Jr., 1999. "Biological Classification / Taxonomy / Systematics /
    Biosystematics : A Penultimate Draft. (non published)."

Leakey, M, and Hay, R., 1979. "Pliocene Footprints in the Laetolil Beds at Laetoli,
    Northern Tanzania." Nature. 278:317-323.

LeGros Clark, W., 1960 (1959). The Antecedents of Man. Edinburgh University Press.

Edinburgh.

Lovejoy,C.O., 1988. "Evolution of Human Walking." Scientific American. 259:118-125.

Olson, T., 1981. "Basicranial Morphology of the Extant Hominoids and Pliocene

Hominids: The New Material from the Hadar Formation, Ethiopia, and its

Significance in Early Human Evolution and Taxonomy." In Aspects of Human

Evolution. Stringer, C, (ed). Taylor and Francis. London.

Pilbeam, D., 1972. The Ascent of Man: An Introduction to Human Evolution. The

Macmillan Series in Physical Anthropology. New York: Macmillan Publishing Co., Inc.

Reader, J., 1988 (1989). <u>Missing Links, 2nd edition.</u> Viking. NY.

Simons, E., 1972. <u>Primate Evolution</u>. MacMillan. NY.

Simons, E., 1989. "Human Origins." <u>Science</u>. 245:1343-1350

Wolpoff, M.H., 1996. <u>Human Evolution: 1996-1997 edition.</u> McGraw Hill. NY.

Early Speech

Dunbar,R., 1993. "Co-Evolution of Neocortex Size, Group Size, and Language in Humans."Behavioral and Brain Sciences. May 2nd 1993.

Gibbons, M.F., Jr., 1977. "The Anatomy of Speech." <u>Science</u>. 195:774-775.

Gibbons, M.F., Jr., 1981. "An Evolution of Speech.". <u>Biology Digest.</u>

Lieberman, P., 1975. <u>On the Origins of Language.</u> MacMillan. NY.

Lieberman, P., The Evolution of Speech. MacMillan. NY.

Lieberman, P., Crelin, E., and Klatt., 1972. "Phoenetic Ability, and Related Anatomy of the Newborn and Adult Human, Neanderthal Man and the Chimpanzee." American Anthropologist. 74:287-307.

Linden, E., 1974. Apes, Men, and language. Penguin. NY.

MacNeilage, P.,1983. The Production of Speech. Springer. NY.

Negus, V.E., 1949. The Comparative Anatomy and Physiology of the Larynx. London: Heinemann.

Walker, S.F. 1988. "Language, Handedness, and the Larynx." Behavioural and Brain Sciences, 11, 731-2.

Zuckerman, S., 1932. The Social Life of Monkeys and Apes. Routeledge and Kegan Paul. London.

Phylogenetics and Cladistics

Deetz, J., 1968. "The Inference of Residence and Descent Rules from Archeological Data." In:New Perspectives in Archeology. Binford, S.R., and Binford, L.R., (eds). Aldine. NY.

Gibbons, M.F., Jr., 1999. "Biological Classification / Taxonomy / Systematics / Biosystematics : A Penultimate Draft. (non published)."

Hill, J.N.,and Evans, R.K., 1972. "A Model for Classification and Typology." In: Models in Archeology. Clarke, D.L., (ed). Methuen. London.

Landau, M., 1984. "Human Evolution as Narrative." American Scientist. 72:262-267.

Tattersall, I., and Eldredge, N., 1977. "Fact, Theory, and Fantasy in Human Paleontology." American Scientist. 65:204-211.

Taxonomy and Variation

Andrews, 1978. "A Revision of the Hominoidea of East Africa." <u>Bulletin of the British Museum of Natural History (Geology)</u>. 30(2):85-224.

Fleagle, J. G., and Mittermeier, R.A., . 1980. "Locomotor Behavior, Body Size, and Comparative Ecology of Seven Surinam Monkeys." <u>American Journal of Physical Anthropology</u>. 52: 301-314.

Mayr, E. 1988. <u>Toward a New Philosophy of Biology: Observations of an Evolutionist</u> Belknap Press/Harvard University Press.

Sexual Dimorphism

Andrews, P., and Martin, L., 1987. "Cladistic Relationships of Extant and Fossil Hominoids." <u>Journal of Human Evolution</u>. 16:101-118.

Delson, E., 1981. "Paleoanthropology:Pliocene and Pleistocene Human Evolution." <u>Paleobiology.</u> 7:298-305.

Fleagle, J. G., and Mittermeier, R.A., . 1980. "Locomotor Behavior, Body Size, and Comparative Ecology of Seven Surinam Monkeys." <u>American Journal of Physical Anthropology</u>. 52: 301-314.

Kay, R., and Simons, E., 1980. "Sexual Dimorphism in Early Australopithecines." <u>Nature.</u> 287:328-330.

Ontogenic Variation

Howell, F.C., 1977. "Hominoidea.". In: <u>Evolution of African Mammals</u>. Maglio, V., and Cooke, H.B.S., (eds). Harvard University Press.

Johanson, D, and Taieb, M., 1976. "Plio-Pleistocene hominid Discoveries in Hadar, Ethiopia." <u>Nature.</u> 260:293-297.

Johanson, D., and White, T., 1979. "A Systematic Assesment of Early African Hominoids." <u>Science</u>. 203:321-330.

McHenry, H., 1975. "Fossils and the Mosaic Nature of Human Evolution."

Science.190:425-431.

Skelton, R., McHenry, H., and Drawhorn, G., 1986. "Phylogenetic Analysis of Early Hominids." <u>Current Anthropology</u>. 27:21-43.

Racial Variation

Bray, W., 1986. "The Paleoindian Debate." <u>Nature</u>. 332:107.

Cuppy, W., 1931. <u>How to Tell Your Friends From the Apes.</u> Horace Liveright. NY.

Smith, F., and Spencer, F., (eds)., 1984. <u>The Origins of Modern Humans: A World Survey of the Fossil Evidence.</u> Alan Liss. NY.

Trinkaus, E., and Howells, W., 1979. "The Neanderthals." <u>Scientific American</u>. 24:118-134.

Polymorphic Variation

Fleagle, J. G., and Mittermeier, R.A., . 1980. "Locomotor Behavior, Body Size, and Comparative Ecology of Seven Surinam Monkeys." <u>American Journal of Physical Anthropology</u>. 52: 301-314.

Hill, J.N., 1968. "Broken K Pueblo:Patterns of Form and Function." IN: <u>New Perspectives in Archeology</u>. Binford, S.R., and Binford, L.R., (eds). Aldine. NY.

Landau, M., 1984. "Human Evolution as Narrative." <u>American Scientist</u>. 72: 262-267.

Pathological Variation

Deetz, J., 1968. "The Inference of Residence and Descent Rules from Archeological Data." In: New Perspectives in Archeology. Binford, S.R., and Binford, L.R., (eds). Aldine. NY.

Hill, J.N., 1968. "Broken K Pueblo:Patterns of Form and Function." IN: <u>New Perspectives in Archeology</u>. Binford, S.R., and Binford, L.R., (eds). Aldine. NY.

Tattersall, I., and Eldredge, N., 1977. "Fact, Theory, and Fantasy in Human

Paleontology." <u>American Scientist</u>. 65:204-211.

Trinkaus, E., and Howells, W., 1979. "The Neanderthals." <u>Scientific American</u>. 24:118-134.

Straus, W., Jr., and Cave, A., 1957. "Pathology and the Posture of Neanderthal Man." <u>Quarterly Review of Biology</u>. 32:348-363.

Temporal Variation

Deetz, J., 1968. "The Inference of Residence and Descent Rules from Archeological Data." In: New Perspectives in Archeology. Binford, S.R., and Binford, L.R., (eds). Aldine. NY.

Hill, J.N., 1968. "Broken K Pueblo:Patterns of Form and Function." IN: <u>New Perspectives in Archeology</u>. Binford, S.R., and Binford, L.R., (eds). Aldine. NY.

Tattersall, I., and Eldredge, N., 1977. "Fact, Theory, and Fantasy in Human Paleontology." <u>American Scientist</u>. 65:204-211.

Taphonomy

Deetz, J., 1968. "The Inference of Residence and Descent Rules from Archeological Data." In: New Perspectives in Archeology. Binford, S.R., and Binford, L.R., (eds). Aldine. NY.

Hill, J.N., 1968. "Broken K Pueblo:Patterns of Form and Function." IN: <u>New Perspectives in Archeology</u>. Binford, S.R., and Binford, L.R., (eds). Aldine. NY.

Tattersall, I., and Eldredge, N., 1977. "Fact, Theory, and Fantasy in Human Paleontology." <u>American Scientist</u>. 65:204-211.

Pliocene

Dart, R.A., 1925. "Australopithecus africanus: The Man-Ape of South Africa." <u>Nature</u>. 115:195-199.

Fleagle, J.G., 1988. <u>Primate Adaptation and Evolution</u>. Academic Press. San Diego, CA.

Fleagle, J.G., 1999. <u>Primate Adaptation and Evolution.</u> Academic Press. San Diego, CA.

Grine, F., 1988. <u>Evolutionary History of the "Robust" Australopithecines.</u> Gruyter: New York

Apes and Humans

Delson, E., 1985. <u>Ancestors: The Hard Evidence.</u> Alan Liss. NY.

Fleagle, J.G., 1988. <u>Primate Adaptation and Evolution.</u> Academic Press. San Diego, CA.

Grine, F., 1988. <u>Evolutionary History of the "Robust" Australopithecines.</u> Gruyter: New York

Simons, E.L., and Pilbeam, D.R., 1965. "Some Problems of Hominid Classification. (a broad overview from the mid 1960's. Origin and diversification of hominids)." <u>American Scientist.</u> 53:237-246.

Simons, E., 1972. <u>Primate Evolution.</u> MacMillan. NY.

White, T.D. and Folkens, P.A., 1991. <u>Human Osteology.</u> Academic Press.CA.

Wolpoff, M., 1996. <u>Human Evolution: 1996-1997 edition.</u> McGraw-Hill. NY.

Australopithecus Africanus

Broom, R., 1936. "The Pleistocene Anthropoid Apes of South Africa." <u>Nature.</u> 142:377-379.

Broom, R., 1947. "Further Remains of the Sterkfontein Ape-Man, *Plesianthropus.*" <u>Nature.</u> 160:430-431.

Dart, R.A., 1925. "Australopithecus africanus: The Man-Ape of South Africa." <u>Nature.</u> 115:195-199.

Fleagle, J.G., 1988. <u>Primate Adaptation and Evolution.</u> Academic Press. San Diego, CA.

Fleagle, J.G., 1999. <u>Primate Adaptation and Evolution.</u> Academic Press. San Diego, CA.

Grine, F., 1988. <u>Evolutionary History of the "Robust" Australopithecines.</u> Gruyter: New York

Howell, F.C., 1977. "Hominidae." In: Evolution of African Mammals. Magio, V.,

And Cooke, H.B.S., (eds). Cambridge and Harvard University Press.

McHenry, H., 1975. "Fossils and the Mosaic Nature of Human Evolution." Science.

190:425-431.

Beynon, A. D. and Wood, B. A., 1986. "Variations in Enamel Thickness and Structure in East African Hominids." American Journal of Physical Anthropology. 70 (2): 177-193.

The Africanus Pattern

Dart, R.A., 1925. "Australopithecus africanus: The Man-Ape of South Africa." Nature. 115:195-199.

Fleagle, J.G., 1988. Primate Adaptation and Evolution. Academic Press. San Diego, CA.

Fleagle, J.G., 1999. Primate Adaptation and Evolution. Academic Press. San Diego, CA.

Grine, F., 1988. Evolutionary History of the "Robust" Australopithecines. Gruyter: New York

Howell, F.C., 1977. "Hominidae." In: Evolution of African Mammals. Magio, V.,

And Cooke, H.B.S., (eds). Cambridge and Harvard University Press.

Beynon, A. D. and Wood, B. A., 1986. "Variations in Enamel Thickness and Structure in East African Hominids." American Journal of Physical Anthropology. 70 (2): 177-193.

The *robustus* Pattern

Beynon, A. D. and Wood, B. A., 1986. "Variations in Enamel Thickness and Structure in East African Hominids." American Journal of Physical Anthropology. 70 (2): 177-193.

Fleagle, J.G., 1988. Primate Adaptation and Evolution. Academic Press. San Diego, CA.

Fleagle, J.G., 1999. Primate Adaptation and Evolution. Academic Press. San Diego, CA.

Grine, F., 1988. Evolutionary History of the "Robust" Australopithecines. Gruyter: New York

Howell, F.C., 1977. "Hominidae." In: Evolution of African Mammals. Magio, V.,

And Cooke, H.B.S., (eds). Cambridge and Harvard University Press.

Lovejoy, C.O. 1981 "The Origin of Man." Science. 211(4479):341-350.

Lovejoy, C.O. 1981 "Models of Human Evolution". Science. 217:304-306.

White, T. D. 1988. "The Comparative Biology of 'Robust' Australopithecus: Clues from Context." Evolutionary History of the "Robust" Australopithecines.. Grine, F.E., (ed). Gruyter. NY

Homo Habilis

Fleagle, J.G., 1988. Primate Adaptation and Evolution. Academic Press. San Diego, CA.

Fleagle, J.G., 1999. Primate Adaptation and Evolution. Academic Press. San Diego, CA

Johansen, D., Masao, F., Eck, G., White, T., Kimbel, W., Asfaw, B., Manega, P., Ndessokia, P., and Suwa, G., 1987. "New Partial Skeleton of *Homo habilis* from Olduvai Gorge, Tanzania." Nature. 327: 205-209.

Leakey, L.S.B., Tobias,P.V., and Napier, J.R. 1964. "A new species of genus Homo from Olduvai Gorge." Nature. 202: 7-9.

Tobias, P.V. 1967. Olduvai Gorge, Volume 2. The Cranium of Australopithecus (Zinjanthropus) boisei. Cambridge University Press.

How Many Genera

Fleagle, J.G., 1999. Primate Adaptation and Evolution. Academic Press. San Diego, CA

Gibbons, M.F., Jr., 1981. "The Human Ancestry." Biology Digest.

Grine, F.E., Jungers, W., and Schulz, J., 1996. "Phenetic affinities among early Homo crania from East and South Africa." Journal of Human Evolution. 30(3):189-225.

Hutchinson, G.E., 1965. The Ecological Theatre and the Evolutionary Play. Yale University Press.

Robinson, J.T. 1972. Early Hominid Posture and Locomotion. University of Chicago Press.

Homo erectus

Brauer, G, and Smith , F., 1992.Continuity or Replacement: Controversies in Homo sapien Evolution. Balkema.

Brown, F., Harris, J., Leakey, R., and Walker, A., 1985. "Early Homo Erectus Skeleton From West Lake Turkana, Kenya." Nature. 316:788-792

Ciochon, R.L., and larick, R., 1996. "The African Emergence and Early Asian Dispersals of the Genus Homo: Early Hominids Began Leaving Africa Almost One Million Years Before Previously Thought." American Scientist. 84:6.

Fleagle, J.G., 1988. <u>Primate Adaptation and Evolution.</u> Academic Press. San Diego, CA.

Fleagle, J.G., 1999. <u>Primate Adaptation and Evolution.</u> Academic Press. San Diego, CA.

Gibbons, M.F., Jr., 1981. "The Human Ancestry." <u>Biology Digest</u>.

Rightmire, G.P., 1990. <u>The Evolution of Homo erectus: Comparative Anatomical Studies of an Extinct Human Species.</u> Cambridge University Press.

Rightmire, G.P., 1992. "Homo erectus: Ancestor or Evolutionary Side Branch?" <u>Evolutionary Anthropology</u> 1:43-49.

Ruff, C., 1993. " Climatic Adaptation and Hominid Evolution:The Thermoregulatory Imperative." <u>Evolutionary Anthropology.</u> 2:53-60.

Swisher, C.C., III., Curtis, G.H., and Jacob, T., 1994. "Age of Earliest Known Hominids in Java, Indonesia." <u>Science.</u> 263:1118-1121.

Walker, A., 1993. "The Origin of the Genus Homo." In: <u>The Origin and Evolution of Humans and Humanness.</u> Rasmussen, D.T., (ed). Jones and Bartlett Publisher. MA.

Wood, B., 1992 "Origin and Evolution of the Genus Homo." <u>Nature.</u> 355:783 - 790.

Homo sapiens

Brauer, G, and Smith , F., 1992.<u>Continuity or Replacement: Controversies in Homo sapien Evolution.</u> Balkema.

Gibbons, M.F., Jr., 1981. "The Human Ancestry." <u>Biology Digest</u>.

Walker, A., 1993. "The Origin of the Genus Homo." In: <u>The Origin and Evolution of Humans and Humanness.</u> Rasmussen, D.T., (ed). Jones and Bartlett Publisher. MA.

Phylogeny

Gibbons, M.F., Jr., 1981. "The Human Ancestry." <u>Biology Digest</u>.

Frayer, D.W., Wolpoff, A.G., Smith, F.H., and Pope, G.G., 1993. "Theories of Modern Human Origins: The Paleontological Ttest." <u>American Anthropologist</u>. 95(1):14 - 50.

Wolpoff, M., 1995. <u>Human Evolution. 1996 edition</u>. McGraw-Hill. NY.

The Neandertals

Aiello, L.C.1993. "The Fossil Evidence for Modern Human Origins in
    Africa: A revised review." <u>American Anthropologist</u>. 95:73-99.

Ambrose, S.1998. "Late Pleistocene Human Population Bottlenecks, Volcanic Winter,
    the Differentiation of Modern Humans." <u>Journal of Human Evolution</u>. 34:623-51.

Bachrach, L.K., Hastie, T., Wang, M.C., Narasimhan, B., and Marcos, R. 1984.
    "Bone mineral acquisition in healthy Asian, Hispanic, Black, and Caucasian   Youth:
A longitudinal study." <u>Journal of Clinical Endocrine Metabolism</u>.
    84:4702-4712.

Balliard, T.1997. "Modern human origins: The faunal remains." <u>Bioessays</u>.
25(11):50-54.

Balliard, T., Sizer, F.S., and Lim, S. 1998. "Neandertals and other archaic *Homo*
    sapiens." <u>American Journal of Physical Anthropology.</u> 70:241-250.

Battisti, M., Chung, K.F., and Donnelly, L.E. 1996. "In Support of the Out-of-Africa
    Theory of Human Evolution." <u>American Journal of Physical Anthropology</u>.
    65:243-252.

Battisti, M., Donnelly, L.E., and Hamilton, M.N. 1991. "Reconstruction of African
    Human Populations." <u>Paleogeography, Paleoclimatology, Paleoecology</u>. 623-651.

Balieu, L. 1999. "Neuroactive Neurosteroids: Dehydroepiandrosterone (DHEA) and
    DHEA sulphate." <u>Acta Pediatricics, supplement</u>. 88:78-80.

Balieu,L., Howard, J.M., and Dutz, G. 1999. "Dehydroepiandrosterone: Metabolic and Developmental Effects." Clinical Endocrine Metabolism. 77:1123-1125.

Binford, L.R. 1985. "Human ancestors: Changing Views of Their Behavior." Journal of Physical Anthropological Archeology. 4:292-327.

Binford,L.R. 1981. "Food Production in the Stone Age." Journal of Physical Anthropological Archeology. 11:221-237.

Bloom, J., Bronage,T., and Schrenk, F. 1979. "Lower Pleistocene Hominids." Journal of Human Evolution. 28:109-114.

Bobyleva, T. R., Cassoli,P.F., Mallengini, F., Piperno,M., and Solano, A. 1993. "The Neandertal Debate." Current Anthropology. 25:301-330.

Bonatti, R.J., Chase, P.G. 1994. "Glacial Puzzles." American Journal of Physical Anthropology. 88(2):525-562.

Brauer, G. 1992. "Modern Human Dispersal Patterns." American Journal of Physical Anthropology. 83(1):249-253.

Brauer, G., and Stringer, C. 1989. "Models of Human Evolution: The Current Debate." Journal of Human Evolution. 17:744-747.

Brauer, G., and Stringer, C. 1997. "Models, Polarization, and Perspectives on Modern Human Origins." In: Conceptual Issues in Modern Human Origins Research. Clark, G.A., and Willermet., C.M.,(eds). Aldine de Gruyter. NY.

Brown, F. 1980. "Early Homo erectus from West Lake Turkana, Kenya." Nature. 316:788-792.

Bunn, H.T. 1981. "Archeological Evidence for Meat-Eating Plio-Pleistocene Hominids from Koobi Fora and Olduvai Gorge." Nature. 291:574-577.

Burrows, J., Alfonzo, D., Maslov, A., and Meyeres, A. 1997. "Dietary Reconstruction in the Upper Pleistocene." Antiquity. 70(279):552-563.

Butzer, E., Johnson, E., and Klein, R.G. 1975. "Behavior and Modern Human Origins." Journal of World Prehistory. 9:167-198.

Cann,R.L.1992. "African Origins of Modern Humans." Scientific American. 266(4):66-69.

Cann, R.L., and Lum, K.L. 1991 "Human Evolution." Endevour. 21(2):91-93.

Cann, R.L., Ching, C., Lum, K.L., and Richards, O. 1994. "Polynesian Mitochondrial DNAs Reveal Three Deep Maternal Lineage Clusters." Human Biology. 66(4):567-574.

Cann, R.L., and Jorde,L.B. 1997. "Mitochondrial DNA Variation." Human Biology. 68(1):1-28.

Cavalli-Sforza, L.L., Prazza, A., Menozzi, P., and Mountain, J. 1988. "Historical Genetic Patterns." Proceedings of the National Academy of Sciences of the USA. 2:6002-6006.

Cavalli-Sforza, L.L., Prazza, A., Menozzi, P. 1998. "The Genetics of Archeology." Proceedings of the National Academy of Sciences of the USA. 4:1420-1427.

Coon, C. 1962. "The Origin of Races." Knopf. NY.

Coop, T. 1989. "Late Pleistocene Paleoecology and Archeology." Journal of Archeological Sciences. 6:235-243.

Chang, N., Li, H.W., and Sadler, L.A. 1993. "Low Nucleotide Diversity in Man." Genetics. 129:513-523.

Davis, K. 1977. "The Ungulate Remains from Kebrara Cave." Journal of Anthropological Research. 47:239-248.

Day, H. 1969. "Stratographical Records of Environmental Change in the Pleistocene." Paleogeography, Paleoclimatology, Paleoecology. 80:25-34.

Dykes, A.H. 1995. "Genetic Affinities." Evolutionary Anthropology. 2:60-67.

Edlund, A., Coop, T., and Sarenthein, R.T. 1995. "The Evolution of Modern Humans During the Upper Pleistocene." Current Anthropology. 26(4):413-419.

Escalante, N., Valladas, H., Joron, J., and Vandermeersch, B.1994. "Out of Africa: New Perspectives on "Eve"." Current Anthropology. 40(3):341-344.

Fellostein, S.1999. "Inference of Human Evolution Through Cladistic Analysis." American Journal of Physical Anthropology. 81:113-124.

Fleming, R. H. 1995. "The Role of Evolutionary Ideas." Choice. 35(6):919-923.

Foley, K. 1994. "Long River Profiles, Tectonism, and Eustosy." Journal of Geogphysical Research. 99:14031-14037.

Foley, K., and Harpending, H.C. 1994. "Quarternary Climatology." Journal of Geophysical Research. 92:1398-1403.

Frayer, R.T.1982. "Demographic Structure of Early Hominid Populations." Journal of Variability and Evolution. 7:3-9.

Frayer, R.T.1984. "Inference in Human Evolution." Journal of Variability and Evolution. 3:32-37.

Frayer, D. 1991. "Mousterian Variability." Journal of Anthropological Research. 47(2):39-44.

Frayer, D., Wolpoff, M., and Thorne, A.G.1992. "Hominids of the Pleistocene." Journal of Variability and Evolution. 11(3):47-53.

Frayer, D., and Thorne, A.G., and Pope,G.1993. "Pleistocene Habitats." Journal of Anthropological Research. 17(1):21-23.

Frayer, D., Thorne, A.G. and Smith,F.H. 1994. "Theories of Modern Human Origins: the Paleontological Test." Journal of Human Variability. 7(9):76-82.

Gargett, S.1989. "The Evolution of Modern Humans:Recent Evidence from Southwest Asia." Journal of Human Evolution. 12:731-742.

Gilman, M.1983. "Continuity in Human Evolution: Bodies, Brains, and the Role of Variability." American Journal of Physical Anthropology. 40:329-341.

Goodson, N. Nei, M, and Graur, D.1991. "How We Got Here." Theoretical Population Biology. 21:132-134.

Goodson, N. Nei, M, and Dangas, W.R. 1996. "Divergence Time and Population Size in the Lineage Leading to Modern Humans." Theoretical Population Biology. 48:198-221.

Gould, S.J. 1987. The Flamingo's Smile: Reflections in Natural History. W.W. Norton and Company. NY.

Gould, S.T.1992. "What is a Species?" Discover. December:40-45.

Graves, P. 1991. "New Models and Metaphors for the Neandertal Debate." Current Anthropology. 32(5):737-739.

Halverson, J.S.1987. "Pleistocene Gravels in the Lower Loye." Journal of Geographic

Research. 99:14031-14050.

Harding, H.M.1997. "The Lower Paleolithic of the Neareast." Journal of World
Prehistory. 8:211-265.

Harold, C.L., Kollett, R., and Chang, N. 1997. "Lineage Hypotheses in Hominid
Evolution." Evolutionary Theory. 11:31-36.

Hawkes, M.T., and Walker, A. 2000. "Foraging for Faunal Resources." British
Archeological Reports. International Series. 163:51-60.

Hayes, D.J., Molnar, P., and Gardner, T. 1991. "Stratigraphy and Ecology in the Middle
Pleistocene." Journal of Quarternary Science. 12:539-545.

Holliday, T.W. 1997. "Evidence of Cold Adaptation in European Neandertals."
Journal of Variability and Evolution. 22(7):245-259.

Howard, J.M. 1999. "Dehyroepiandrosterone, Melatonin, and Testosterone in
Human Evolution." Steroids. 37:345-352.

Howard, J.M. 2000. "Androgens and Evolution." Steroids. 9:78-80.

Howell, F.C. 1951. "The Place of Neanderthal Man in Human Evolution." American
Journal of Physical Anthropology. 9:379-416.

Howell, F.C. 1952. "Pleistocene Glacial Ecology and the Evolution of "Classic"
Neanderthal Man." Southwestern Journal of Anthropology. 8:337-410.

Howell, F.C. 1957. "The Evolutionary Significance of Variation and Varieties of
"Neanderthal" Man." Quarterly Review of Biology. 32:330-347.

Ishikawa, S., Ishikawa, A., Yoh, K., Tanaka, H., and Fujiwara, M. 1999.
" Osteoporosis in Male and Female Leprosy Patients." Calcification
Tissues International. 64:144-147.

Jarnick, M., Labudda, D., Stoneking, M., Korab-Laskowska, M., Tishkoff, S., Batzer, M.,
Modiano, D., and Scozzari, R. 1998. "Genetic Structure of the Ancestral
Populations of Modern Humans." Journal of Molecular Evolution. 47(2):146-155.

Jin, T., Bromage, T., and Schrenk, F. 1995. "Biogeographical and Climatic Basis for a
Narrative of Hominid Evolution." Journal of Human Evolution. 28:109-114.

Johnson, T.C. 1989. "The Neanderthals and Population as a Prime Mover." Current
Anthropology. 30(4):534-536.

Jorde, R. 1991. "Genetic Structure of Early Hominids." <u>Journal of Molecular Evolution.</u> 276(2):129-131.

Jorde, R., Thorne, A.G., and Smith, F.H. 1998. "Neanderthals Debated." <u>Journal of Variability and Evolution.</u> 65:109-113.

Jurmain, R., Nelson, H., and Kilgore, L. 1995. <u>Essentials of Physical Anthropology, 2nd edition.</u> West publishing Co. St Paul, MN.

Klein, R.G. 1989. <u>The Human career: Human Biological and Cultural Origins, 2nd ed.</u> University of Chicago Press. Chicago and London.

Kollett, R., Wang, H., and Raheel, M. 1993. "Evidence, Mechanisms, and Controversy." <u>American Journal of Physical Anthropology.</u> 75:193-202.

Krings, M.A., Stone, A., Schmitz, W., and Kraninitzki, H., Stoneking, M., and Paabo, S. 1997. "Neandertal DNA Sequences and the Origin of Modern Humans." <u>Cell.</u> 90(1):19-30.

Kuzawa, P. 1998. "Plesistocene Land-Sea Correlations." <u>Earth Science Reviews.</u> 13:307-311.

Lahr, M.M. 1994. "Gondal Hormone Influences on the Emergence of Cortical Function in Non-Human Primates." <u>Behavioral Neuroscience.</u> 103:1287-1295.

Lahr, M.M., and Foley, R.A. 1998. "Towards a Theory of Modern Human Origins: Geography, Demography, and Diversity in Recent Human Evolution." <u>Yearbook of Physical Anthropology.</u> 41:137-145.

Leonard, W.J. 1994. "No Severe Bottleneck During Human Evolution." <u>American Journal of Human Genetics.</u> 48:383-389.

Lieberman, P., and Crelin, E., 1971. "On the Speech of Neanderthal." <u>Linguistic Inquiry</u>. 2:203-333.

Lieberman, P., Crelin, E., and Klatt, D.H., 1972. "Phonetic Ability and Related Anatomy of the Newborn and Adult Human, Neanderthal Man and the Chimpanzee." <u>American Anthropologist.</u> 74:287-307.

Lieberman, L.S. 1982. "Normal and Abnormal Sexual Dimorphic Patterns of Growth and Development." In: <u>Sexual Dimorphism in *Homo sapiens*, a Question of size.</u> Hall, R.L. (ed). Praeger. NY

Linfield, L., Stute, M., and Simpson, H.J. 1999. "Evolution of Variation." <u>Journal of</u>

Variability and Evolution. 22(7):245-251.

Lum, Y.M. 1989. "The Late Quarternary Fauna." Quarternary Research. 27:239-243.

Marieb, E.N. 1996. Human Anatomy and Physiology, 2nd edition. Addison Welsey
Longman, Incorporated.

Martel, T.E., Bunn, H.T., and Kroll, E.M. 1998. "Ancestrial Behavior." Journal of
Anthropological Research. 33:493-502.

Melentes, F.T., and Roe, D. 1979. "Human Evolution." Bioessays. 18:945-954.

Mellars, P. , Kunzig, R., and Lewontin, R. 1987. "Evolution Revisited." Journal of
Human Evolution. 32:43-46.

Mellars, P. 1988. "The Neandetal Problem Continued." Current Anthropology.
40(3):341.

Milton, E. 1987. "The Myth of "Eve."" Evolutionary Anthropology. 4:53-63.

Milton, E. 1999. "The Fate of the Neanderthals: Between about 30,000 and 40,000 years
ago, the Neanderthals in Europe were replaced by populations of behaviorally
and biologically modern humans. What happened during that period?" Journal
of Human Evolution. 39:179-181.

Nadler, W.H., Burrows, W., and Ryder, O.A. 1980. "Populations of the Pleistocene."
Human Molecular Genetics. 4:1485-1487.

Offner, P.J., Moore, E.E., and Biffele, W.L. 1999. "Male Gender is a Risk Factor for
Major Infections After Surgery." Archives of Surgery. 134:935-938.

Ozasa, H., and Gould, K.G. 1982. "Demonstration and Characterization of Estrogen
Receptor Chimpanzee Sex Skin: Correlation Between Nuclear Receptor Levels
and Degree of Swelling." Endocrinology. 111:125-131.

Park, M.A., 1999. Biological Anthropology, 2nd edition. Mayfield Publishing Company.
Mountain View, CA.

Parsons, H. 1997. "Earliest Human Occupation in Central Asia." Current Anthropology.
36:337-346.

Payne, A.K., Binford, L.R., Zilhao, J., and Marean, C. 1983. "Paleoclimatic Changes and
Faunal Remains." Journal of Archeological Sciences. 22:495-503.

Perel, E., Davis, S., and Killinger, D.W. 1981. "Androgen Metabolism in Male and

Female Breast Tissue." Steroids. 37:345-352.

Perry, H.M. 3rd , Horowitz, M., Morley, J.E., Fleming, S., Jensen, J., Caccione, P., Miller, D.K., Kaiser, F.E., and Sundarum, M. 1996. "Aging and Bone Metabolism in African American and Caucasian Women." Journal of Clinical Endocrine Metabolism. 81:1108-1117.

Potts, R.B. 1988. Early Hominid Activities at Olduvai. Aldinede Gruyter. NY.

Potts, R.B. 1996. "Evolution and Climate Variability." Science. August 16:922-23.

Quintana-Murci, L., Semino, O., Bandelt, H., Paparino, G., McElreavey, K., and Santachiara-Benerecetti, A.S. 1999. "Genetic Evidence of and Early Exit of *Homo sapiens* from Africa Through Eastern Africa." Nature Genetics. 23:437-441.

Rak, Y. 1993. "Remains from Amud Cave." American Scientist. 58(2):60-69.

Rak, Y., Kimbel, W.H., and Hovers, E. 1994. "A Neandertal Infant from Amud Cave Israel. Journal of Human Evolution. 26:313-324.

Relethford, J.H., and Harpending, H.C. 1994. "Ancient Differences in Population Size Can Mimic a Recent African Origin of Modern Humans." Current Anthropology. 36(4):667-674.

Rightmire,G.P. 1990. The Evolution of *Homo erectus*, Comparative Anatomical Studies of an Extinct Species. Cambridge University Press.

Rightmire, G.P.1995. "Putting Neandertals Back in Our Family Tree." Science. 252.

Roberts, M.A. 1997. "Population Growth." Molecular Biological Evolution. 9:552-569.

Roberts, M.A. 1999. "Human Biological Origins." Nature Genetics. 23:437-441.

Rogers, R. 1998. "Phylogeny in Evolution." New Scientist. 128(1743):33-35.

Ruvolo,M.E., Carretero, J.M., and Arsuaga, J.L. 1993. "Late Pleistocene Archeology." Current Anthropology. 3(1):347-353.

Sarenthein, L. 1978. "The Faunal Remains." American Anthropologist. 94:306-307.

Sarrel, P., Dobay, B.J., and Willta, B. 1998. "Estrogen and Estrogen-Androgen Replacement in Post-Menopausal Women Dissatisfied with Estrogen

Only Therapy: Sexual Behavior and Neuroendocrine Responses."
Reproductive Medicine. 43:847-856.

Shakelton, M., Kroll, E.M., and Klein, R.G. 1977. "Re-Analysis of Faunal
Assemblages." Journal of Archeological Science. 13:515-518.

Simpson,G.G. 1945. The Principles of Classification and a Classification of Mammals.
Bulletin of the American Museum of Natural History. 85(ixvi).

Simpson, G.G. 1949. The Meaning of Evolution: A Study of the History of Life and
of its Significance for Man. Yale University Press. CT.

Smith, J.M. 1976. "Behavior and Evolution." Scientific American. 239(3):176-181.

Smith, S.L. and Harold, F.B. 1990. "Evolution Rates in Plants." Bioessays. 11:32-35.

Smith, F.H. "Bodies and Brains in Hominid Evolution." Journal of Variability and
Evolution.12:13-17.

Smith, F.H., Steudel, K., and Norgan, N.G. 1997. "Early Hominid Body Size." American
Journal of Physical Anthropology. 52:63-70.

Stein, P., and Rowe,B. Physical Anthropology, 5th edition. McGraw Hill, New York.

Stoneking, M, and Cann, R. 1989. "African Origins of Human Mitochondrial DNA." In:
The Human Revolution. Mellars, A., and Stringer, C. editors. 17-30. Zdinburg
University Press.

Stoneking, M. 1993. "A Case for "Eve."" Scientific American. 232(6)71-72.

Stoneking, M. , and Cann, R. 1998. "Mitochondrial DNA and Human Evolution."
Nature. 325:31-36.

Stringer,C.B., Aldhous, S., and Pettitt, P. 1979. "European Neanderthals." Natural
History. 92:6-9.

Stringer, C.B., Grun, R., and Frayer, D. 1985. "Hominids of the Pleistocene." Scientific
American. 64(3):44-45.

Stringer, C.B., and Andrews, P. 1988. "Genetic and Fossil Evidence for the Origin of
Modern Humans." Science. 239(1743):33-38.

Stringer, C.B., and Brauer, G. 1993. "Early Asian Hominid Populations." Antiquity.
57(247):943-946.

Tagliaferro, A.R., Davis, J., Truchon, S. and Van Hamilton, N. 1986. "Effects of Dehydroepiansrosterone Acetate on Metabolism, Body Weight and Composition of Male and Female Rats." Journal of Nutrition. 116:1977-1183.

Tattersall, I.1986. "Species Recognition in Human Paleontology." Journal of Human Evolution. 15:165-175.

Tchernov, P. 1994. "Biological Change in the Late Pleistocene." Proceedings of The National Academy of Sciences, USA. 88(1):3942-3946.

Templeton,A.R. 1993. "Perspectives in Human Biology." Evolutionary Anthropology. 2(2):53-56.

Tortura, G.J., Funke, B.R., and Case, C.L. 1996. Microbiology, an Introduction, 5th edition. 425-441. Benjamin Cummings. Redwood City, CA.

Treisman, R. 1995. "The Multiregional and Single Origin Hypotheses of the Evolution of Modern Man: A reconciliation." Journal of Theoretical Biology. 173:23-29.

Trinkaus, E. 1986. "The Neanderthals and Modern Human Origins." Annual Review of Anthropology. 15:193-218.

Trinkaus, E. 1989. "Documenting the Origin of Modern Humans." In: The Emergence of Modern Humans: Biocultural Evolution in the Late Pleistocene. Trinkaus, E. (ed). Cambridge University Press.

Vigilant, L., Stoneking, M., Harpending, H., Hawkes, K., and Wilson, A.C. 1991. "African Poulations and the Evolution of Human Mitochondrial DNA." Science. 153:1503-1507.

Vandermeersch, B.1985. "The Origin of Neanderthals." In: The Ancestors: The hard Evidence. 306-309. Delson, E. (ed). A.R. Liss. New York.

Verdonck, A., Gaethofs, M., Carels, C., and deZegher, F.1999. "Effect of Low-Dose Testosterone Treatment on Craniofacial Growth in Boys with Delayed Puberty." European Journal of Orthodontics. 21:137-142.

Wallace,A.R.1992. "Anthropology and Race." American Journal of Physical Anthropology. 73:33-40.

Webster, D.B.1987. "Rates of Evolution." <u>American Journal of Physical Anthropology.</u> 32:99-109.

Wickings, E.J., and Nieschlag, E., 1999. "Seasonality in Endocrine and Exocrine Testicular Function of the Adult Rhesus Monkey (*Macaca mulatta*)." <u>International Journal of Andrology.</u> 3:87-104.

Wiehmann, M.W., Zellweger, R., DeMaso, C.M., Ayala, A. and Chaudry, I.H. 1996. "Mechanisms of Immunosuppression in Males Following Trauma-Hemorrage: Critical role of Testosterone." <u>Archives of Surgery.</u> 131. 1186-1191.

Wiley, E. 1998. "Calibration of Upper Pliocene-Lower Pleistocene Fossil Events with Oxygen Isotope Stratigraphy." <u>Paleooceanography.</u> 8:88-100.

Wilson, A., and Cann, R. 1992. "The Recent African Genesis of Humans." <u>Scientific American.</u> 266(4):68-73.

Wolpoff, M., Thorne, A., and Lawn, R. 1991. "The Case Against Eve. (theory that humans originate from a single African woman refuted by new evidence fossils)." <u>New Scientist</u>. 130(1774):37-41.

Wolpoff, M., and Thorne, A.G. 1992. "The Multiregional Evolution of Humans." <u>American Scientist</u>. 266(4):76-81.

Wolpoff, M., Frayer, D.W., Thorne, A.G., Geoffrey, F.H., and Pop, G. 1994. "Getting it Straight. (response to article by Chris Stringer and Gunter Brauer)."<u>American Anthropologist.</u> 96(2):424-439.

Wolpoff, M., and Caspari, R., 1997. <u>Race and Human Evolution: A fatal attraction.</u> Simon and Schuster.

Wood, B.A., and Straits, D.S. 1991. "Early Hominid Biogeography." <u>Proceedings of The National Academy of Sciences USA</u>. 9:9196-9200.

Zietkiewicz, B., Turner, A., and Suwa, G. 1998. "Human Migration in the Pleistocene." <u>Indo-Pacific Prehistory Association Bulletin</u>. 14:22-26.

<u>Comparative Anatomy and Ostoeology of Extant Primates: Additional Identification</u>

Techniques

Aiello, L., and Dean, C., 1990. An Introduction to Human Evolutionary Anaotomy. Academic Press. London.

Aiello, L., and Wheeler, P., 1995. "The Expensive Tissue Hypothesis: the brain and Digestive system in human and primate evolution". Current Anthropology. 36:199-221.

Andrews, P, and Groves, C.P., 1976. "Gibbons and Brachiation." Gibbons and Siamang. Rumbaugh, D. M., ed., IV, S. Karger, Basel.

Ankel-Simons, F., 1983. A Survey of Living Primates and Their Anatomy. Macmillan. NY.

Ankel-Simons, F., 1999. Primate Anatomy, 2nd edition. Academic Press. NY.

Basmajian, J.V., 1985. Biofeedback: Principles and Practice. Williams and Wilkins. NY.

Basmajian, J. V., and De Luca, C., 1985. Muscles Alive: Their Functions Revealed by Electromyography. 5th ed.Williams and Wilkins. Baltimore, MD.

Bennet, K.A., 1993. A Field Guide for Human Skeletal Identification, 2nd edition. C.C. Thomas. Springfield, IL.

Biegert, J. , 1963. "The Evaluation of the Skull, Hands , and Feet for Primate Taxonomy." In: Classification and Human Evolution. Washburn, S.L., editor. Aldine Atherton. Chicago, IL.

Biggerstaff, R.H., 1977. "The Biology of the Human Chin". In: Orofacial Growth and Development. Albert, T., Dahlberg, A., and Graber, T.M.,(eds.), Hague- Mouton.

Brothwell, Brothwell, D. 1981. Digging Up Bones. Oxford University Press. London.

Bruckner, J.S., 1992. "The Human Subtalar Joint: A theme on Variations". PhD Dissertation. Indiana University.

Cahill, D.R., 1965. "The Anatomy and Function of the Contents of the Human Tarsal Sinus and Canal." Anatomical Record. 153:1-18.

Campbell, B., 1985. Human Evolution, 3rd edition. Aldine Publishing. New York, NY.

Cartmill, M., 1975. Primate Origins. Burgess Press. Minneapolis, MN.

Ciochon, R. L., and Fleagle, J.G., 1987. Primate Evolution and Human Origins. Aldine De Grutyner. New York, NY.

Demes, B, and Creel, M, 1988. "Bite Force, Diet, and Cranial Morphology of Fossil Hominids". Journal of Human Evolution 17:657-670.

DuBrul. E.L., and Sicher, H., 1954. The Adaptive Chin. Charles Thomas. Springfield.

Elefson, J.O., 1967. "A Natural History of Gibbons in the Malay Penunsula." Ph D Thesis. University of California at Berkeley.

Fleagle, J.G., 1985. "Size and Adaptations in Primates" In: Size and Scaling in Primate Biology. Jungers, W.L., editor. Plenum Press, London.

Fleagle, J.G. 1988. Primate Adaptation and Evolution. Academic Press. San Diego, CA.

Fleagle, J.G., Janson, G.C., and Reeds, K., 1999. Primate Communities. Cambridge University Press. New York, NY.

Fox, G.J., 1972. "Some Comparisons Between Siamang and Gibbon Behavior". Folia Primatologica. 18:122-139.

Groves, C.P., 1989. The Biology of Race. University of California Press. Berkeley.

Hartman, C.G., and Straus, Jr., W.L., 1933. The Anatomy of the Rhesus Monkey. Williams and Wilkins. Baltimore, MD.

Haxton, H.A., 1947. "Muscles of the Pelvic Limb". Anatomical Record. 98:337-346.

Heaton M. J. 1979. "Cranial Anatomy of Primitive Captorhinid Reptiles from the Late Pennsylvanian and Early Permian Oklahoma and Texas." Bulletin of the Oklahoma Geological Survey , 127: 1-84.

Hershkovitz, P., 1977. Living New World Monkeys (Platyrrhini) with an Introduction to the Primates. Volume One. University of Chicago Press. Chicago, IL.

Hill, W.C.O., 1972. Evolutionary Biology of the Primates. Academic Press. New York, NY.

Hill, W.C.O., 1974, (1953-1974). Primates: Comparative Anatomy and Taxonomy. Edinburgh University Press. Edinburgh, and Wiley-Interscience, New York,NY.

Hillson, S W, 1996. <u>Dental Anthropology</u>. Cambridge University Press. Cambridge.

Hylander, W.L. 1985, "Mandibular Function and Bimechanical Stress and

Scaling." <u>American Zoologist</u>. 25: 315-330.

Jenkins, F.A., Jr., 1974. <u>Primate Locomotion</u>. Academic Press. New York, London. Jungers,

W.L., 1985. "Body Size and Scaling in Limb Proportions in Primates". In:

<u>Size and Scaling in Primate Biology</u>. Jungers, W.L., editor. Plenum Press.

London.

Jones, F.W.,1916. <u>Arboreal Man</u>. Edward Arnold , London.

Kavanaugh, M., 1983. <u>A Complete Guide to Monkeys, Apes, and Other Primates</u>.

Viking Press. New York, NY.

Kay, R.F, and Fleagle, J.G., 1994. <u>Anthropoid Origins. Advances in Primatology</u>.

Plenum Publishing. New York, NY.

Kennedy, K.A.R., and Chiment, J., 1992. "Racial Identification in the Context of

Prehistoric-Historic Biological Continua: Examples from South Asia."

<u>Social Science and Medicine</u>. 34(2):119-123.

LeGros Clark, W., 1960, (1959). <u>The Antecedents of Man</u>. Edinburgh University

Press. Edinburgh.

LeGros Clark, W., 1961. <u>The History of the Primates</u>. University of Chicago Press.

Chicago, IL.

Lieberman,D.E.,2000. "Ontogeny, Homology, and Phylogeny in the Hominid

Craniofacial Skeleton: The problem of the browridge." In: <u>Development, Growth</u>   <u>and</u>

<u>Evolution: Implications for the study of hominid skeletal evolution</u>.

O'Higgins,P., and Cohn, M.,(eds.) Academic Press. London.

Lohman, T.G., Roche, A.F., and Martorell, R., 1988. <u>Anthropometric Standardization</u>

<u>Reference Manual.</u> Human Kinetics Books. Champaign, IL.

Mann, R., and Imman, V.T., 1964. "Phasic Axtivity of Intrinsic Muscles of the Foot."

<u>Journal of Bone and Joint Surgery</u>. 46A(3):469-481.

Marieb, E.N., 1997. <u>Human Anatomy and Physiology, 4th edition.</u> Benjamin Cummings.

San Francisco, CA.

Marieb, E.N., 2000. Human Anatomy and Physiology, 5th edition. Benjamin Cummings. San Francisco, CA.

Miles, A.E.W., 1962. "Assessment of Age from the Dentition." Proceedings of the Royal Society of Medicine. 51:1057-1050.

Molnar, S., 1998. Human Variation: Races, Types, and Ethnic Groups. Prentice Hall.

Morbeck, Preuschoft, H., and Gomberg, N., 1979. Environment, Behavior, and Morphology: Dynamic interactions in primates. Fischer. New York, NY.

Morton, D.J., 1924. "Evolution of the Longitudinal Arch of the human Foot." Journal of Bone and Joint Surgery. 6:56-90.

Napier, J.R., and Davis, P.R., 1959. "The Forelimb Skeleton and Associated Remains of Pro-Consul africanus." Fossil Mammals of Africa. 16.

Napier, J.R., and Napier, P.H., 1967. A Handbook of Living Primates. Academic Press. London.

Nelson, D.A., 2001. "Ethnic Differences in Bone Mass and Architecture." American Journal of Physical Anthropology. Annual. 112.

Nowak, R.M., 1991. Walker's Mammals of the World. John Hopkins University Press Baltimore, MD.

Osborn H. F. 1903. "On the Primary Divison of the Reptilia into Two Sub- Classes, Synapsida and Diapsida." Science 17: 275-276.

Owen, R., 1959. On the Classification and Geographic Distribution of the Mammalia. Parker. London.

Picq,P.G., 1994. "Craniofacial Size and Proportions and the Functional Significance of the Supraorbital Region in Primates." Zeitschrift für Morphologie und Anthropologie 80(1):51-63.

Picq, P.G., and Hylander, 1985, W., 1989. "Endo's Stress Analysis of the Primate Skull and the Functional Significance of the Supraorbital Region". American Journal of Physical Anthropology. 79:393-398.

Pope, G.G., 1992. "Taxonomic Diagnosis and Morphological Definition of Anatomically

Modern Homo Sapiens." <u>American Journal of Physical Anthropology</u>.

    (Supp. 14):133.

Preuschoft, H., and Chivers, D.J., editors, 1993. <u>Hands of Primates</u>. Springer-Verlag.

    New York, NY.

Ravosa, M.J., 1991. "Structural Allometry of the Prosimian Mandibular Corpus and

    Symphysis." <u>Journal of Human Evolution</u> 20:3-20.

Reichs,K.J., 1998. <u>Forensic Osteology: Advances in the Identification of Human</u>

    <u>Remains.</u> C.C. Thomas. Springfield, IL.

Roche, A.F., Heymsfield, S.B., and Lohman, T.G., editors, 1996. <u>Human Body</u>

    <u>Composition. Human Kinetics</u>. Champaign, IL.

Rowe, N., 1996. <u>The Pictorial Guide to the Living Primates</u>. Pognonias Press.

    East Hampton, NY.

Sauer, N.J., 1992. "Forensic Anthropology and the Concept of Race: If Races Don't

    Exist, Why are Forensic Anthropologists so Good at Identifying Them?"

    <u>Social Science and Medicine</u>. 34(2):107.

Savage, R. J. G., 1972. "Review of the Fossil Mammals of Libya". <u>Symposium on the</u>

    <u>Geology of Libya.</u>

Savage, R.J.G., and Long, M.R., 1986. Mammal Evolution, an Illustrated Guide. Facts of

    File Publications. New York, NY.

Schnapf, J.L. , and Schneeweis, D.M., 1999. <u>Electrophysiology of Primate Cone</u>

    <u>Photoreceptors. in Color Vision: From Genes to Perception</u>.

    Cambridge University Press. Cambridge.

Schneeweis, D.M., and Schnapf, J.L. 1999. "Photovoltage of Macaque Cone

    Photoreceptors: Adaptation, noise and kinetics."<u>Journal of Neuroscience</u>

    19:1203-1216.

Schultz A. H., 1936. "Characters Common to Higher Primates and Characters Specific

    for Man." <u>Quarternary Review of Biology</u>. 11: 259-283 and 425-455.

Scott, J., 1963. "Factors Determining Skull Form." <u>Symposia of the Zoological Society of London</u>. 10:127-134.

Stern, J.T., and Susman, R.L., 1973. "Primate Locomotion: some links with evolution and morphology." <u>Primatlogia</u>. IV:11

Swindler, D.R., and Wood, C.D., 1973. <u>An Atlas of Primate Gross Anatomy: Baboon, Chimpanzee and Man</u>. University of Washington Press. Seattle, WA.

Warwick, R., and Williams, P.L., 1973. <u>Gray's Anatomy, 35th Brittish Edition</u>. WB Saunders. Philadelphia, PA.

White, T.D., 1991. <u>Human Osteology</u>. Academic Press. San Diego, CA.

White, T.D., 1999. <u>Human Osteology, 2nd edition</u>.   Academic Press. San Diego, CA.

Conclusion

Bown, T.M., 1976. "Affinities of Teilhardina (Primates, Omomyidae) with Description of a New Species from North America." <u>Folia Primatologica</u>. 25:62-72.

Bown, T.M., and Rose, K.D., 1987. "Patterns of Dental Evolution in Early Euocene Anaptomorphine Primates (Omomyidae) From the Bighorn Basin, Wyoming." <u>Paleontological Society.</u> 23:1-162.

Brown, F., Harris, J., Leakey, R., and Walker, A., 1985. "Early Homo Erectus Skeleton From West Lake Turkana, Kenya." <u>Nature</u>. 316:788-792

Carroll, R.L., 1997. <u>Patterns and Processes of Vertebrate Evolution.</u> Cambridge University Press.

Covert, H.H., 1995. "Locomotor Adaptations of Eocene Primates: Adaptive Diversity

Among the Earliest Prosimians." In: <u>Creatures of the Dark: The Nocturnal -</u>
Prosimians. Alterman, L., Doyle, G.A., and Izard, M.K., (eds.). Plenum Press.

Covert, H.H., 1997. "The Earliest Primate Adaptive Radiations and New Evidence about
Anthropoid Origins." In: <u>Biological Anthropology: The State of the Science.</u> Boaz,
N.T., and Wolfe, L.D., (eds.). Oregon State University Press.

Darwin, C., 1859. <u>Origin of the Species.</u> (a facsimile of the first edition). Harvard
University Press.

Delson, E., 1984. "Cercopithecid Biochronology of the African Plio-Pleistocene:
Correlation Among Eastern and Southern Hominid-Bearing Localities." <u>Cour.
Forsch.-Inst. Senckenberg.</u> 69:199-218.

Fleagle, J.G., 1988. <u>Primate Adaptation and Evolution.</u> Academic Press. San Diego, CA.

Fleagle, J.G., 1999. <u>Primate Adaptation and Evolution.</u> Academic Press. San Diego, CA.

Fleagle, J.G., and Kay, R.F., 1994. Anthropoid Origins (Advances in Primatology).
Plenum, N.Y.

Gibbons, M.F., Jr., 1981. "Marsupials –Their Place in the Natural History of Mammals."
<u>Biology Digest</u>. 8:2.

Gibbons, M.F., Jr., 1981. "The Human Ancestry." <u>Biology Digest</u>.

Howells, W., 1997. Getting Here: <u>The Story of Human Evolution</u>. Compass Press.
Washington, D.C.

Huxley, T.H., 1872. <u>Lay Sermons, Addresses, and Reviews.</u> Appleton.NY.

Huxley, T.H., 1872. <u>Lessons in Elementary Physiology</u>. MacMillan. London.

Huxley, T.H., 1872. <u>A Manual of the Anatomy of Vertebrated Animals.</u> Appleton. NY.

Kay, R.F., Ross, C., and williams, B.A., 1997. "Anthropoid Origins." Science. 275:797-
804.

LeGros Clark, W., 1926. "On The Anatomy of the Pentailed Shrew *Ptilocercus lowii*."

Proceedings of the Zoological Society of London. 46:1179-11809.

LeGros Clark, W., 1934. Early Forunners of Man. London.

LeGros Clark, W., 1960 (1959). The Antecedents of Man. Edinburgh University Press.

Edinburgh.

McHenry, H., 1975. "Fossils and the Mosaic Nature of Human Evolution." Science.

190:425-431.

Mirvart, G.J., 1873. Man and Apes. London.

Wolpoff, M., 1984. "Evolution in *Homo erectus*: The Question of Stasis." Paleobiology.

10(4):389-406.